Integrated Product Development

IFS

INTEGRATED PRODUCT DEVELOPMENT

M. M. Andreasen
and
L. Hein

IFS (Publications) Ltd, UK

Springer-Verlag
Berlin · Heidelberg · New York
London · Paris · Tokyo

M. M. Andreasen/L. Hein
Institute for Product Development (IPU)
The Technical University of Denmark
Building 423
DK-2800 Lyngby
Denmark

658.57
A55i

British Library Cataloguing in Publication Data
Andreasen, M. Myrup
    Integrated product development.
    1. New products
    I. Title   II. Hein, Lars
    III. Integreret produktudvikling. *English*
    658.5'75   HD69.N4

ISBN 0-948507-21-7 IFS (Publications) Ltd
ISBN 3-540-16679-3 Springer-Verlag Berlin
ISBN 0-387-16679-3 Springer-Verlag New York

©1987 **IFS (Publications) Ltd,** 35-39 High Street, Kempston,
                          Bedford MK42 7BT, UK
    Springer-Verlag, Berlin Heidelberg New York
                 London Paris Tokyo

This work is protected by copyright. The rights covered by this are reserved, in particular those of translating, reprinting, radio broadcasting, reproduction by photo-mechanical or similar means as well as the storage and evaluation in data processing installations even if only extracts are used. Should individual copies for commercial purposes be made with written consent of the publishers then a remittance shall be given to the publishers in accordance with §54, Para 2, of the copyright law. The publishers will provide information on the amount of this remittance.

This book was first published in 1985 by Jernets Arbejdsgiverforening, Copenhagen, under the title: 'Integreret Produktudvikling'.
Translated by Robin Sharp

Phototypeset by Systemset, Stotfold, Bedfordshire.
Printed and bound by Short Run Press Ltd, Exeter.

Foreword

Design is central to manufacturing industry. That is even more true in today's climate of high technology and consumer expectancy. Design has a direct impact on a company's bottom line. It determines not only the functional performance and reliability but also the cost of the product. But, no longer can design be treated in isolation. It impacts on every part of a manufacturing company's business and equally those same parts impact on design.

This lesson, recognised by the Japanese for a long time, is now being learnt by companies in the West. This global strategic approach can be best described as product development and through involving the whole business becomes integrated product development (IPD).

IPD can be defined as the process of taking a product through the many interlinked stages of a company's business from concept to sales and installation. The concept stage involves marketing, innovation, entrepreneuralism, and so on. Design must reflect the marketing needs whilst ensuring that the product can be manufactured at the right cost and quality. Manufacture has to make the product as designed and in the right quantities at the right time. Sales need to sell the product produced by manufacture – not what it would like to sell! The result of IPD should be a successful business.

The authors of this book are experts on the subject. They are both Danish – and Denmark like all the other Scandinavian countries has a fresh outlook on design. Form and appearance are important but so are functionality and value-for-money. This is reflected in Scandinavian products. In addition, the authors have for many years appreciated the criticality of design and the need to consider design as an integral part of business activity. Professor Andreasen, in particular, teaches engineering design and design methodology as well as being a consultant in design and automation to many large companies in Europe. The Technical University of Denmark, where he practices, considers design to be a general discipline unlike many other European institutes where design is just another subject in engineering. All of this undoubtedly adds to the value of the book and provides the reader with not only the basics of integrated product development but also the means whereby he/she can implement the philosophy into their business.

The realisation that design is an integral part of product development is only just occurring to many companies, but it is an awakening process; and IPD is being recognised as a discipline in its own right by many companies in Europe and the USA. There is, therefore, a demand for education in the subject and this book can provide the vehicle for an educational programme. It is likely to be primarily of value to senior management responsible for the strategical operations of their company but engineering designers and those responsible for production will benefit from its contents. I recommend it as a contribution towards the improvement of business efficiency.

R. Armstrong
Director-General, the Production Engineering Research
Association (PERA) of Great Britain and Northern Ireland.

Contents

Preface

The subject of this book is product development within manufacturing industry. The significance of product development for the survival and competitiveness of industrial companies is well-known. But product development doesn't operate under the conditions it should do, it is not carried out with sufficient professionalism and it does not exploit its possibilities sufficiently.

The message of this book is that a company must come together, must integrate itself, in order to carry out product development, because this activity, in contrast to most others that take place within a company, cannot tolerate being split up.

The book is primarily intended for top managers within industry, for project leaders, and for project participants, but we firmly believe that teachers and students following technical or business-oriented courses will benefit from reading it as well.

Integrated Product Development has arisen as part of a research project of the same name, carried out by the Engineering Design Group at the Institute for Product Development (IPU) in Denmark. The project was supported by the Danish Board of Technology, the Thomas B. Thrige Foundation, the V. Aa. Jeppesen Memorial Foundation, the Institute for Engineering Design at the Technical University of Denmark, and IPU. In addition, a number of development, consultancy and teaching activities for industrial firms provided their experience which we have used as part of the basis of this book.

We are professionally indebted to a number of persons who have given us insight or ideas through their publications or in discussions. In particular we are indebted to Prof. Fredy Olsson of Lunds University, who has previously formulated the integration concept, Prof. J. Eekels of Delft University, Hans Mikkelsen of Konsulentgruppen A/S, Knud Valbjoern of Danfoss, and many others. Mention must also be made of the project's reference group, and of the other members of the project group at IPU.

Mogens Myrup Andreasen
Lars Hein

To the reader!

You will get more out of this book – just like any other book – if you are aware of the authors' intentions when writing it.

The book is about product development. It is not about what sort of product to develop. Nor is it about how to discover what sort of product to develop. But it deals with how to make the tiny fraction of your activities which are concerned with setting objectives, organisation, planning, management, leadership, choice of methods, and so on, more professional and efficient, so that the remaining (the major) part of your activities gives good results.

It can be looked upon as being a textbook, but it is not really sufficiently comprehensive, unbiased or broad enough to deserve such a title. Our attitude has been pragmatic rather than scientific, and we have drawn on sources, results and points of view which are particularly suited to conditions in manufacturing industry.

The book should be regarded more as a collection of essays, in which the subject of product development is reconsidered from a particularly basic point of view. This is, that our organisations, which are split up according to function and speciality, and our procedures, which are heavily schematised, run contrary to the nature of product development –and are out of step with the demands being made on companies today.

We concentrate on procedures, activities and methods. One of the reasons for this is to counteract the widespread belief that once you've set up the right organisation then everything else just happens by itself. Our experience is the opposite. Sensible product development can take place within almost any organisation, if the employees just know what to do.

Large and small companies

The reader should try to use this book as a mirror, in other words to hold up the book's points of view and its provocative assertions against his or her own situation, so as perhaps to be able to recognise new possibilities. It is especially important that readers from small concerns don't assume in advance that integration is a problem which is only of importance for large companies. This book offers the small company numerous useful guidelines for what to do, if you wish to engage in product development.

The authors

Mogens Myrup Andreasen, Ph.D. has been with the Laboratory for Engineering Design, The Technical University of Denmark, since 1966, where he is now associated professor of Mechanical Engineering. He received his Ph.D. in 1980 from The University of Lund in Sweden. He is involved in the teaching of Engineering Design and Design Methodology and he has won international acclaim for his research and publications in these fields. He is a consultant to IPU and to numerous Danish companies.

Lars Hein has been with The Institute for Product Development (IPU) since 1975, where he is now manager of the Product & Manufacturing Development Department. His work has covered management of some of IPU's major projects for Danish industry and he serves as a consultant to a number of Danish companies. During the years 1980–84, he was project manager on the research projects which pioneered IPU's work on integrated product development, and he is now manager of IPU's main project in this field.

Why integrated product development?

SUMMARY

Product development is a very complex activity, and it is tempting to divide it up into:

- tasks related to the market, to the product and to its production
- tasks in the short term, project-oriented tasks and tasks related to long-term strategy
- tasks related to management, to project management, and to project execution
- the individual projects which are in hand at one time
- the individual development activities (of which product development is one) which are in hand at one time

Several troublesome problems are caused by this differentiation, and can be solved by using an integrated approach.

What problems are we thinking of here? In this chapter, we shall describe the problems which are experienced, or indeed caused, at the level of the management, the project leader and the project team-member, and those problems which are 'nobody's'! How these problems are related to the task of integration are then shown.

1.1 Five areas of integration

The term 'Integrated Product Development' arises from the idea that product development cannot be carried out in the best possible way if it is allowed to disintegrate into different areas of specialisation, areas of activity or areas of responsibility. Let us take a closer look at what integration involves:

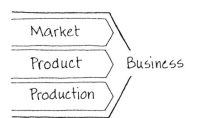

1. Three tasks are to be carried out at one time

The aim of product development is the creation of good business for the company. Good business is the result of using the market, the product and its production to maximum advantage. If this task is shared among the departments concerned – marketing/sales, development/design, and production – there is a considerable risk that the company will fail to maximise its potential and will drift off course.

- Keep your balance
- Move forward
- Reach your target

2. Three time-frames are to be taken into account

The result of product development is running production and sales. These activities require great care and attention to priorities, but they often divert our attention and our resources from the longer-term task of creating new products and – to an even greater extent – from the task of planning and controlling product development in accordance with a long-term strategy.

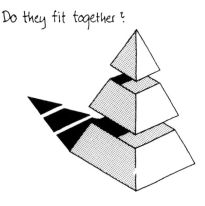

Do they fit together?

3. Three levels of activity have a common objective

People from different levels in the company are involved in setting a strategy, in leading product development projects and in performing the practical tasks. Serious difficulties may arise in creating agreement between aims, means and results on these three levels.

A collision course?

4. Controlled interplay between product development projects

A company usually has quite a number of development projects in progress at the same time: some of them major, with long-term significance, together with many smaller ones which ensure short-term profits, and which are necessary in order to satisfy customer requirements, for example. If there is no controlled interplay between these projects, with establishment of priorities, then the small ones will gain greater priority for themselves and disturb the larger ones, so that the total result will be confusion. Projects must therefore be integrated.

5. Controlled interplay between development activities

Many elements of a dynamic company must be in a constant state of development, so that the company can adjust itself to the demands made on it from outside, and to the objectives set by the management. At a high level, we therefore speak of organisational, market, product and production development. At a lower level, we speak of quality control, financial control, stock control, sales, packaging, advertising and competitor analysis as areas of development. In addition, we have the company's service functions and functions which help to preserve the level of competence within the company, such as the materials workshop, experimental workshop, logistics, electronic data processing (EDP), inspection, patenting and so on.

How can we ensure – in a large and complex concern – that these activities will have a positive effect on one another, and that the service-related elements do not become restrictive and merely contribute to costs? And how do we ensure – in a small company – that the right quality-producing activities, which these aspects of the organisation reflect, actually get looked after in a project carried out by a modest number of participants, perhaps by just a single person?

Controlled interplay

1.2 Problems

It is not difficult to produce a long list of significant problems; it is much more difficult to provide convincing solutions to them. Nevertheless, we shall try to elucidate the intention of this book by presenting a series of problems which are closely related to lack of integration or to 'disintegrated' thinking and problem solving.

The Executive

Nobody

The Project Leader

The Team-member

We have chosen to relate the problems to four types of people in the development hierarchy, namely: the Executive (the board of directors, product committee, head of development, proprietor or other people with a strategic, long-term responsibility for product development); the Project Leader (the person responsible for the project, head of department, product developer, designer, i.e. a person who bears the responsibility for carrying out a development project, either in its entirety or through certain phases), the Team Member (designer, stylist, estimator, or employee within the production preparation, sales, marketing, quality control, or EDP departments, who takes part in some task as part of a development project); and 'Nobody', by which we mean anybody in the company who, as a result of lack of attention, lack of knowledge or incorrect choice of priorities, fails to carry out necessary tasks.

3

The list of problems given below is a product of the authors' activities as consultants for Danish industry. Of course, not all these problems appear in all the companies which they are acquainted with, but they seem to be characteristic enough, judged by the frequency with which they are met.

The Executive's problems

- "It's too expensive"

Compared with other costs, product development is expensive and the budgets never hold.

- "The efficiency of product development is too low"
- "We get too little out of product development, in spite of..."

Perhaps the company is not at all 'geared up' to product development: development is based on the structure of the organisation; small and short-term tasks get higher priority than larger ones; those persons who should take part in development have other primary tasks and objectives; responsibility is passed on from one person to another; subsidiary results are optimised instead of optimising the business as a whole...

- "We take too long to develop things"

Many projects are started too late, the decision-making process is too complex, and the degree of risk accepted is too small ("We should just look into..."). The management takes too long when deciding to start a project; on the other hand, the early phases are then often carried out too poorly and too hastily, with the result that delays are introduced in the later phases of the project.

- "We have no control over quality"

The marketing department is seldom satisfied with the result of the project, and when sales actually start there can be serious problems with maintaining quality. Problems with quality must be solved in the design phase, in a close collaboration with the marketing and production departments, starting from the basic specification, and with the aim of solving problems rather than having a showdown within the organisation.

- "Production gives us problems with costs and deadlines"
- "We need to modernise our production techniques"

When new products are being developed, things are in a state of flux, and it doesn't cost extra to solve problems correctly. Top-level goal-oriented development of the production must be coordinated with the development of the product and proper design of the products is crucial for determining the costs and controllability of the production.

- "We produce too many/too few variations"

At the same time! Variations are necessary for renewal, and to adapt to the needs of the customer, but they are not conducive to rational production. Mastery of the problem of how to choose a product range is a difficult exercise in collaboration, and requires precise, long-term targets.

- "We are being pressed by our competitors"
- "Too many of our products are failures"

The products don't sell. Couldn't we have foreseen this? Who took an interest in the potential market during the course of the project? Who took an interest in our competitors? Perhaps we aimed too low and took too small a risk.

- "I've said it time and time again, that..."
- "I'm usually pretty well in touch..."

The Executive has vision, and keeps up with developments, as regards management science, new methods and new 'schools'. But at the bottom of the company hierarchy people don't understand what he's aiming at, so they settle for their own lesser objectives. And they feel themselves understimulated and poorly motivated, due to lack of insight into the overall picture.

The Project Leader's problems

- "The project plans aren't being kept to"

The project leaders are often caught between management and their project teams. They become prey to priority conflicts, lack of resources and ill-defined project planning, and their colleagues have to work in a fragmented situation, following up old projects at the same time as they deal with a whole series of current ones.

- "The management doesn't commit itself"
- "The management is beating about the bush"

A project which cuts across the path of other activities means that resources and sovereignty have to be given up to the project, and this presupposes that people within the management and in many other areas commit themselves to working on the project's premises. It is much easier not to do so, or just to deal with minor problems – and indeed this is what tends to happen.

- "I get told too little"

The setting in motion of a development project often takes place after long and difficult consideration, supported by investigations, small experiments, discussions, a bit of consultancy activity, a few more investigations, and so on. When the project is finally started, the target can clearly be seen ahead, and everyone is in a hurry to get going. But the project leader doesn't get much out of this preliminary work, and is badly informed about what implications the project has for the company.

- "I hope it doesn't go wrong"
- "Are we staking enough?"

In these two opposite statements lies partly recognition of the fact that the company doesn't have enough projects in progress for there to be room for failures – or perhaps that the company doesn't accept that development projects can go wrong – and partly a doubt about what is being staked. Is enough being done about this? Are our requirements strong enough? The cautious project leader is often successful at planning, but not always successful at achieving results.

- "What do I get out of it?"

The project leader is in an exposed and demanding position. Everybody can see if the result is poor. Plenty of people are standing waiting to point out what went wrong, and it is difficult to rid your reputation of a failure. Faced with all this, the project leader must be prepared to motivate himself to offering his life-blood in a long and demanding effort, where leisure and family life are not taken into consideration, but where complete commitment to the cause is required. Do the rewards adequately compensate him for this?

The Team Member's problems

- "Shortage of time"

There is never time to do anything properly. When the plans are actually ready and tasks are dealt out, rapid realisation of the plans is demanded. We never consider alternative solutions. The jobs come bundled together. Many problems get solved with interim solutions, and the products never get decently worked through.

- "I lack motivation"

Tasks are solved mechanically, there are continual small difficulties which get dealt with piecemeal; I never have my own job so I can finish it properly. We have lots of discussions with the marketing and production people; the specifications keep changing and we have to alter things all the time.

- "I never get told anything"

It is difficult to see any consistency or direction in management decisions. I don't know why we've suddenly taken up Project 12, but I suppose we are trying to compete our German competitor out of the Swedish market. I think it's a bad idea to base things on those Japanese motors, but nobody listens to what I say.

- "There is no use for my creativity"

We have just had an in-house course, but we don't use any of the methods. There isn't the time, as we're behind with respect to the plans, and the production people want to have the drawings altered so they suit another type of plastic material. The project leader doesn't want anything about formulating the problem and looking at alternatives, he wants some concrete suggestions. I worked a bit off my own bat on the new valve, but they weren't interested.

- "The work is poorly paid"
- "The work is not appreciated"

The management should be able to see that we in the design department are of key importance for the company. Now the sales department has got new offices and invested in a new computer, so there's no money for me to go to the Hanover Show this year. I think I'll look for a new job.

'Nobody's' problems

- Nobody solves the problem of communication

People have not been told who has the responsibility for ensuring that employees get the information which will give them an overall view of things, ensure their continued effort toward a common target, and create a basis for their motivation. Instead of this, tons of paper and dozens of rumours are circulated.

- Nobody solves the problem of motivation

Product development demands the utmost of a company and cannot be dealt with by '9 to 5' workers. They have to be motivated to give an extra, well-coordinated and creative effort, so somebody has to solve the problem of how to motivate them. Personal contact and mutual respect among those who take part in a project are here much more powerful instruments than chains of command and minutes from meetings.

- Nobody seriously looks after product development

In firms which are divided into departments according to function, product development becomes institutionalised and follows fixed procedures. Seen from a number of points of view, product development is a revolutionary activity, where each advance requires a standard bearer, and each project is carried through at the expense of other activities. Activities like this will not stand up to being divided, to the sharing of responsibility, to delegation or to compulsory contributions. Instead they require that some people or someone has an overall view, sets things going, keeps control, provides motivation and stops things at the right time.

- Nobody looks after the early stages

It is vital for the company that it has at its disposal a collection of business opportunities, ideas for projects and ideas for products, so that product developments can continually be set going. In many companies, this is not the case: projects arise by chance and in an opportunistic manner, and demands are made for their rapid concretisation and resolution. The search for good business ideas, on which projects can be based, hardly takes place at all.

- Nobody cares about costs

The cost structures within a company look quite different from the functional differentiation of the company. If you seriously want to do something about costs – and realise how things hang together – then things must be dealt with in an integrated manner across the company. Nobody is responsible for the cost of materials or indirect costs.

1.3 Why integration is needed

A general pattern would seem to emerge from the preceding problem situations: the reason for the problems is the division of duties, and the departmentalisation which takes place as the company expands. This departmentalisation may appear as in Fig. 1.1, and is sought justified by the necessary division of labour, the demand for specialisation, the need for service facilities, and so on.

● IDEAL INTEGRATION = ONE PERSON

In 1967, K.G. Zeuthen designed the 'Postbox' copying machine for the company Zeuthen & Aagaard. Zeuthen had from his young days worked on office machines and duplication equipment. Zeuthen was also responsible for the design of the KZ plane and a long series of other exceptional products – and he is still active today.

The basic concept behind the 'Postbox' copying machine was invented by Zeuthen on his own, and is an ideal example of integration, as Zeuthen at one and the same time had a good grasp of the potential market, was an amazing designer, and had considerable experience of production techniques.

– a growing market

The 'Postbox' was created in the early days of copying machines, when there was a considerable market for machines which could make a single copy from an original. The machine was intended to offer a decentralised solution to the copying problem, and was easy to use and to service.

– a good product

The copying machine contains a simple transport system, which moves the original and the copy in synchronism past the optical system. The machine is assembled on a large, welded chassis, incorporating securing jigs for all the components. After assembly, only a single mirror/lens unit needs to be adjusted.

– ease of production

The chassis is a good basic unit for assembly line production, and the many cleverly designed fastening systems make it unnecessary to use tools, and leave no doubt as to where the individual components are to be placed. All components are simple, and the use of cover plates leads to an elegant design.

The Growth of a Company

1.1 Departmentalisation affects the conditions under which the task of product development takes place, because this task touches and involves many activities within the company, including activities which do not have product development as their primary objective. There is thus a tendency for product development activities to fall to pieces, so basic mechanisms are disrupted. Rather crudely, we may say that the following areas are affected:

- The direction of product development
  because of departmental objectives, delegation of critical responsibilities, lack of motivation, reduced priorities and collective decision-making.

- The efficiency of product development
  because of the fragmentation of tasks, reduced concentration, lack of engagement, reduced awareness of the overall situation and of the external conditions (including the competition!), and the effects of an inappropriate organisation and decision-making structure.

- The effort put into product development
  because this task drowns in a host of other tasks, the willingness to take risks is reduced due to the formal organisation, short-term tasks oust long-term ones, the problem of setting priorities cannot be solved merely by using one's own insight, renewal mechanisms become too complex, and so on.

- The professionalism of product development
  because one is trying to make a creative genius with the help of a mediocre team. Specialisation creates little kingdoms with a no man's land between them, people lack respect for the other departments, but also lack insight into the significance of the early stages, and so on.

The ideal company can be thought of as one in which a single person is in charge, and where knowledge of the market, of design and production and of economic mechanisms are collected together in one person, who is also able to make decisions and willing to run a risk. Many important Danish companies have started in this way, have had considerable success, but have then gone through a critical period when they adopted a more divided form of organisation.

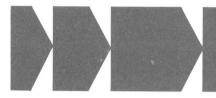

The answer to the problems presented here is not to integrate departments, functions or management tasks, but to introduce integrated procedures, aims, attitudes and methods into product development.

In the Integrated Product Development project, on which this book is based, we therefore worked with the hypothesis that it would be a good idea to create these tools and to use them to counteract the harmful effects of 'disintegration'. The aim of this book is to highlight a number of relationships, tasks and problems which are related to product development, and which can be made more efficient, can be rationalised and can be solved using the tools of integration.

The book is put together in such a way as to cover many important aspects of product development, but it cannot be said to be a complete textbook on product development. Rather it should be considered as a source of inspiration and information for those who already know something about product development.

The book is aimed at 'three levels' within the company: the management, the project leaders and the members of the project teams, but greater attention is given to the tasks and problems of the project leader, because ideally he or she creates and runs the project, sells its results, and acts as the link between the aims of the management and the efforts of the project team. For management, the book points out the role of product development in renewal of the company, and discusses the creation of development projects in accordance with company policies and objectives. For team members, the book defines their role in the development task, and the use of methods which bring about integration.

2

Changed conditions for product development

SUMMARY

The problems illustrated in Chapter 1 are intensified by the great changes taking place in the external conditions under which companies operate, and which make it necessary for every company to exploit product development as the most important tool for renewal and adaptation. The problems are also intensified by the many mistakes we make in the way in which we look at and handle development projects.

This situation is analysed on the basis of the authors' knowledge of Danish industry, and is related to the most important ideas which have to be recognised concerning the nature of product development. The chapter concludes with a selection of courses of action and methods of attack which will be presented in Chapter 3 under the designation of integrated product development.

2.1 Changing conditions

Companies are living under turbulent conditions in which the survival of the individual company doesn't just depend on its ability to sell products here and now, but also on its ability to renew itself. Companies that pay too little attention to renewal, in relation to their competitors and to technological development, will die. So active adaptation is necessary, especially with the help of product development, which is the company's source of energy and which radically influences the company's ability to renew itself with respect to production and sales.

Improved opportunities for trade and economic stagnation in the highly developed industrial countries has brought an upswing in international trade, and therefore increased competition. On the world market compete high-technology countries, countries which have considerably lower wage levels, and countries which have large stable home markets. Product development is exposed to increasing demands with respect to quality and costs.

Fig. 2.1 shows the course followed by technical development, with ever-increasing dynamism and shorter maturation periods. This means that product renewal occurs more often, and, taken together with the increased competition, this means that the life-time of the product becomes shorter. New demands for updating technical knowledge, for know-how, for short development times and for great adaptability to new technology and methods are therefore being made within the area of product development.

No company is today left untouched by the changed attitudes and altered legal requirements with respect to the working environment, to the external environment, to product liability, and to changed conditions as regards energy and material resources. The management of these factors takes place in the central phases of the development task, and introduces new requirements for insight and control over the results of this work.

2.1

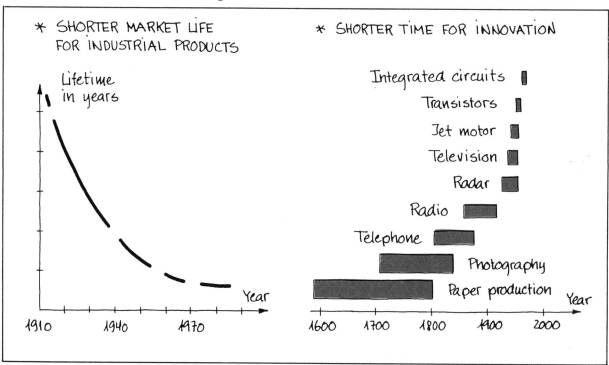

12

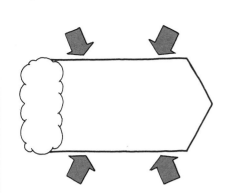

* CAN THE DEVELOPMENT GROUP WITHSTAND THE PRESSURE FROM OUTSIDE?

- Shorter development time requires efficiency/ flexibility/ ready knowledge
- All methods for minimising production costs must be utilised
- Rapid conversion of new knowledge to practical results is required
- Competion/Quality consciousness/ Product responsibility put increased demands on the creative process

2.2

There is no sign that the conditions just referred to will become any easier in the next few years. All the more is therefore required of those responsible for carrying out product development. In view of our unsatisfactory performance, which we shall discuss further in Section 2.2, companies need to concentrate on product development and to strengthen this activity through their choice of strategy, planning and new methods.

The product development departments of companies therefore experience ever-growing external pressure, which can expose the fact that unsuitable ways of going about things are being used. Fig. 2.2 summarises the increased demands which arise from external conditions.

2.2 Influential factors in product development

Within those companies at which this book is aimed, there seem to be a number of factors which influence the activity of product development in quite the wrong direction, and which are at variance with some of the general tenets concerning the nature of product development. The authors base this view on their consultancy and educational activities within Danish industry, on insights into industrial conditions in West Germany, the UK and the USA and on the fact that the pattern is quite different in successful Japan. The findings are outlined below.

● An unfortunate, sequential course of events is followed

When a company grows to a certain size, division of labour and specialisation occur in the areas of marketing, development, design, production planning, production and sales. This is to ensure the greatest possible level of competence and knowledge in these areas, but at the same time it means that a development project may move round from one department to another. Delegation of tasks can be critical, because the newcomers may have different perceptions of the aim and purpose of the project. There will be a tendency to under-achievement in the individual departments if the overall target is unclear or unknown. Employees often have a hazy understanding of their role in the overall scheme of things, and only a limited knowledge of procedures which can ensure good overall results (Fig. 2.3).

13

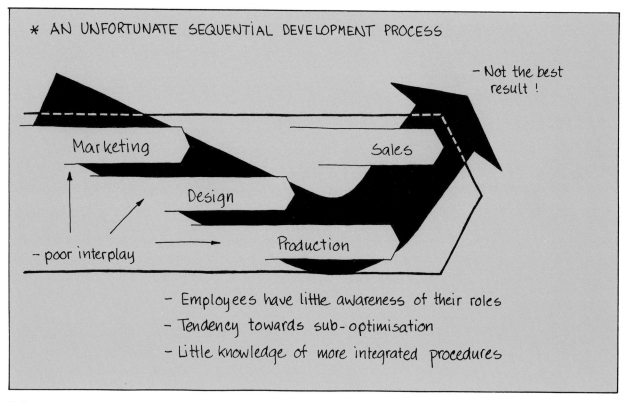

AN UNFORTUNATE SEQUENTIAL DEVELOPMENT PROCESS

Marketing

Sales

Design

Production

– Not the best result !

– poor interplay

- Employees have little awareness of their roles
- Tendency towards sub-optimisation
- Little knowledge of more integrated procedures

2.3

- Market knowledge and design possibilities are badly combined

At an early stage in the project, the designer determines the product's function, its scope, the way in which it will work for the user, its performance, and the range of products of which it will form part. At this stage, the knowledge possessed by the marketing and sales departments is poorly exploited; one assumes that one knows how the market works and puts oneself in the place of both seller and user. The end-use situation is not deliberately defined, but it turns up by chance and on the product's terms. Conversely, the performance of the product, and its various features, are determined without knowledge of what possibilities exist for design or technique. Thus there is only poor interplay between these two areas, caused primarily by unsuitable procedures, as well as ignorance of, and lack of mutual respect between, the departments concerned (Fig. 2.4).

2.4

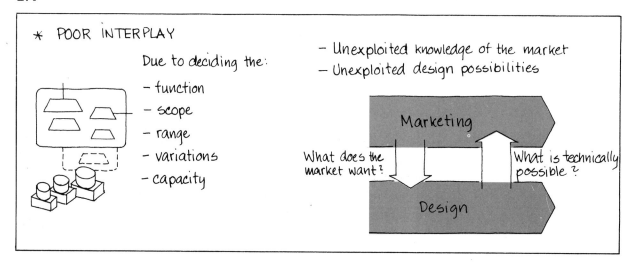

POOR INTERPLAY

Due to deciding the:
- function
- scope
- range
- variations
- capacity

- Unexploited knowledge of the market
- Unexploited design possibilities

Marketing

What does the market want?

What is technically possible?

Design

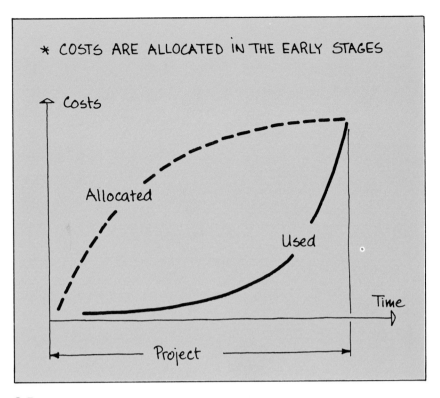

* COSTS ARE ALLOCATED IN THE EARLY STAGES

2.5

● Basic cost relations are neglected

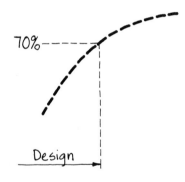

There is a consistent relationship between defrayed and allocated costs in a development project, namely that they do not occur at the same point in time (see Fig. 2.5). Costs are allocated in the early stages; in particular, the production-related costs are for the most part allocated in the design phase – that is to say, up to 70%! This means that the designer should be able to foresee the consequences of his decisions and achieve the best possible results on the basis of a fairly complex economic model, which takes in function, production and, to a certain extent, sales as well. In addition to the early definition of the financial constraints of the project, the possible methods of production are determined at the same stage, as we shall see in the next two statements.

● Limited interplay between design and production methods

At the design stage, when the designer determines the structure of the product, i.e. defines its component parts and the way in which they will be put together, he determines at the same time in broad terms the structure of the production system – that is, the various operations and their sequence – and its dynamic characteristics such as flexibility, dynamics, duration and so on. In so doing he defines the production system's adaptability to sales dynamics. Thus at this stage it is useful to incorporate knowledge and objectives related to production technology, and to make the most of the fact that the product is in a state of flux. Unfortunately, recognition of this fact is poorly utilised in industry; in many companies there is the attitude that one does not interfere with the other departments (Fig. 2.6).

15

● THE DESIGNER'S INFLUENCE

A proposal for structural modifications in a series of trailers from JF Ltd lead to considerable modifications in the production system. By switching over from the use of sections to folded sheet and by replacing a welded axle by a bolt-on axle, it became possible to delay the selection of trailer type and model until later in the production process. This gave increased control and flexibility. How many designers have a grasp of these considerations? [2]

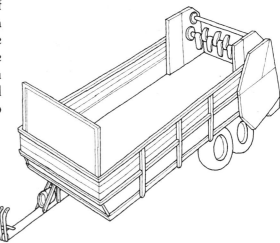

Structural modifications:

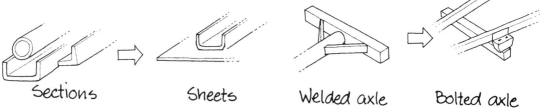

Sections   Sheets   Welded axle   Bolted axle

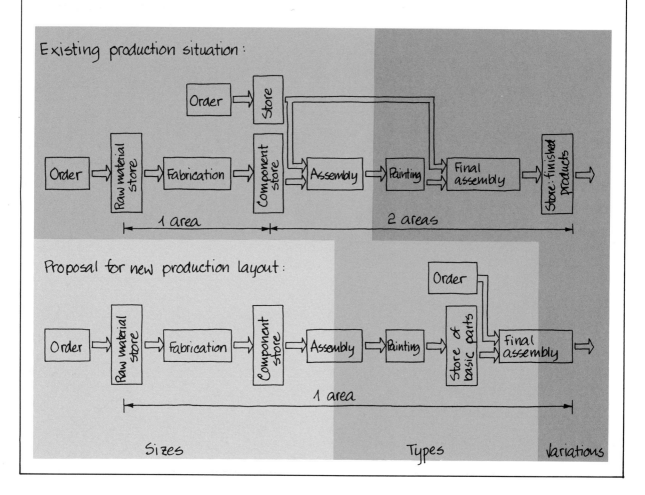

Existing production situation:

Proposal for new production layout:

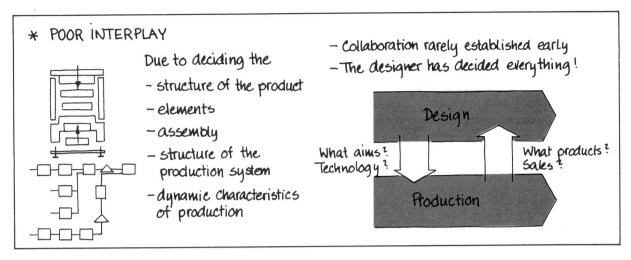

2.6

- Limited interplay between design details and process determination

In the design phase, where the component parts are designed in detail, and the materials, surfaces and tolerances to be used are set, the processes and assembly methods which can be used for production of these parts are to a large extent determined. Detailed drawings thus have a number of technical process decisions built into them, whether the designer realises this or not. At this stage, it is important to channel know-how about processes and assembly techniques into the work of design, thereby influencing the product while the situation is still fluid. People do not generally recognise that once the drawings are finished, the designer has decided everything (Fig. 2.7).

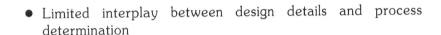

- The early project phases proceed haphazardly and in the wrong way

Development projects should be set going on the basis of a careful analysis of requirements and research into the technical possibilities, and the aims and means of the project should be carefully determined according to the nature of the project. It is our experience that projects start off at a great pace, and very quickly

2.7

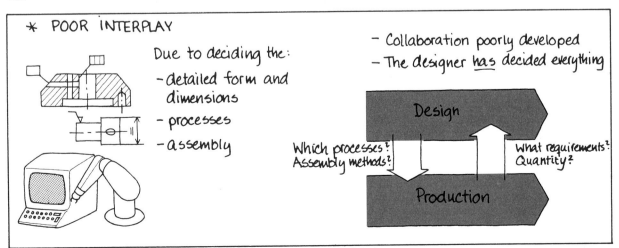

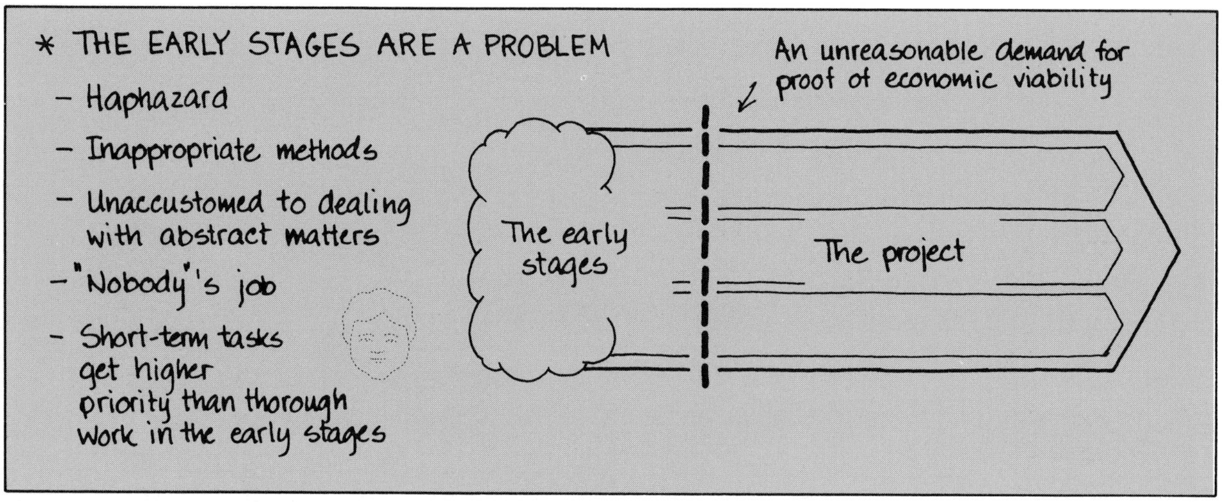
* THE EARLY STAGES ARE A PROBLEM
- Haphazard
- Inappropriate methods
- Unaccustomed to dealing with abstract matters
- "Nobody"'s job
- Short-term tasks get higher priority than thorough work in the early stages
An unreasonable demand for proof of economic viability
The early stages
The project

2.8

become more concrete, so that the lack of proper preliminary analysis becomes apparent later. People are unaccustomed to dealing with abstract matters, the early phases are 'nobody's' tasks, and are left to chance initiatives. The job of consolidating the early phases is long-term, but short-term tasks are often given higher priority.

Management often reinforces this unsuitable way of going about things by insisting on having 'proof' of the economic result demonstrated at the start of the project (Fig. 2.8). This means that one is obliged to make things more concrete at the wrong time, perhaps even to take a great leap forward in the phases of the project. The early phases should have their own, allotted course.

2.3 Courses of action and methods of attack

If we look more closely at the problems and situations which are outlined above, it is not possible immediately to decide where to tackle them. One can justifiably relate these problems to conditions such as the organisation, know-how and methods (professionalism), procedures, plans, project culture, leadership and so on, and one can attempt to choose courses of action within these areas.

It is our opinion that very important problems can be traced back to failure to develop an overall view, and to lack of interplay within the product development project itself. We shall therefore consider product development as the area where the most important effort is required, and will treat failure to take an overall view and the lack of interplay in organisation, knowledge, methods and so on, as secondary matters.

The treatment of our solution, which is introduced under the title of 'Integrated Product Development', will be related to the people introduced in Chapter 1: The Executive, the Project Leader, the Team Member and 'Nobody'. We will, however, lay most stress on the management aspects, both inside and outside development projects, and on the attitudes that should lie behind the activity of product development.

Integrated Product Development is presented in Chapter 3.

Do you, as project leader or designer, feel yourself in this fellow's situation after having read Chapter 2? It is clear that many pointed demands are made on product development in a company, but we are not saying that everything is the designer's responsibility. Many people should be involved at the design stage. (From Kurosawas film version of Macbeth, "Throne of Blood". The general (the actor Mifune) has been impaled on the stockade of the castle by his own soldiers, who have lost confidence in him and regard him as a bringer of bad luck.)

● AREN'T COMPANIES INTEGRATED?

A company may well be said to be integrated when the activities of its many employees lead to results. But in our opinion many companies have concrete ceilings, dividing walls and blinkers, which make collaboration less effective.

Integrated Product Development is an approach to product development whose aim is to create the proper interactions between the separate activities within the company; not like in the mechanical puppet theatre shown here, but as a result of all the employees knowing the aims, roles, methods and tools of integration.

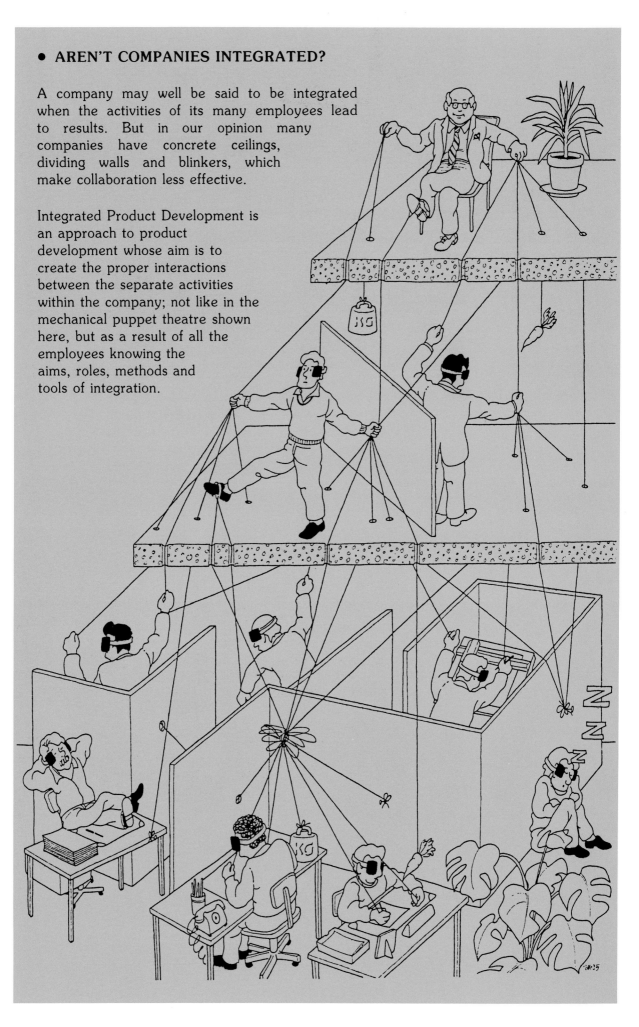

20

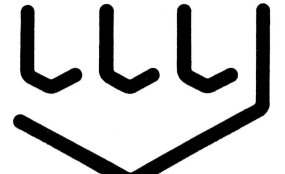

3

What is integrated product development?

SUMMARY

Integrated product development (IPD) is an idealised model for product development, which is integrated in terms of creation of market, product and production, and which clarifies integration between project and management, including the need for continual product planning. Product development should be integrated with other development activities, and contribute to renewal and adaptation within the company. To improve your understanding of the model, we will examine a series of typical development tasks seen with IPD eyes, and will then attempt to clarify the validity of the IPD model.

In short, this chapter introduces the IPD model as the cornerstone of the integration concept; the remainder of the book shows how the integration concept, realised in different ways, can assist in the solution of a series of problems within the company.

3.1 Idealised model for product development

A step-by-step exposition and explanation of an idealised model for product development is now given. In the course of this, the assumptions on which the model is based are explained, and thereby the foundation for the model 'Integrated Product Development' is laid. Later on we shall see that integrated product development is more than just this idealised model.

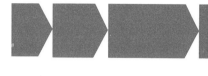

The result of product development should be successful business

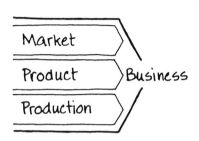

Business is an interaction between the market, the product and its production. This means that the result of carrying out product development is made up of three elements: recognition and creation of a market and establishment of sales outlets; creation of a product which satisfies this market, and which, at the same time, can be produced by the third element: the production system which has been developed for the purpose. All three elements must be present and must make up the best possible overall combination.

We can therefore not carry out product development merely on the basis of a desire to produce an ideal product. No product is so good that it will 'sell itself', or that it doesn't matter what it costs to produce it.

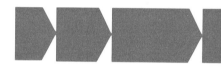

Product development is a creative, multi-disciplinary process

The three main disciplines are marketing/sales, development/design and product development/product establishment/production (Fig. 3.1). So there must be three parallel activities with business as their common objective. These three areas must at one and the same time have their own professional life and content, and be organised with a common leadership and set of goals.

Marketing   Sales

Development / Engineering design

Development/Establishment/Production

* MANY DISCIPLINES !

3.1

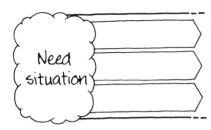

Product development starts with a need

Irrespective of whether a development project is based on radically new ideas or will merely make a small improvement to an existing product, the sole reason for the existence of the product is that it must meet a recognised need. We have therefore chosen to allow our model of product development to start with a cloud symbolising the existence of a need – that diffuse, difficult to grasp situation that can be summed up by saying that 'someone' has a need. It may be consumers with a need for foodstuffs, manufacturers with a need for production equipment, assembly plants with a need for components, and so on. The heart of the matter is to start from a recognised need and to carry it through to the achievement of successful business. What happens during this activity?

Product development is an iterative process...

By iterative is meant that we gradually approach our goal through a number of cycles in which corrections are made. The iterative nature of projects is a consequence of the fact that we do not possess any methods which can take us directly from a problem to its solution. So we must primarily utilise trial-and-error methods, which may involve small or large iterative loops. The loops can be so large that the project is abandoned: the result was far removed from what was expected. But we must also carry out the task of development out as an iteration or by backwards steps, even in the case of quite small details, such as the choice of materials or the shaping of a single component.

... so we need to map out our course

3.2

If the course of progress is not mapped out, we will not have the right conditions for working in an iterative manner. The first rough outline of this map is shown in Fig. 3.2.

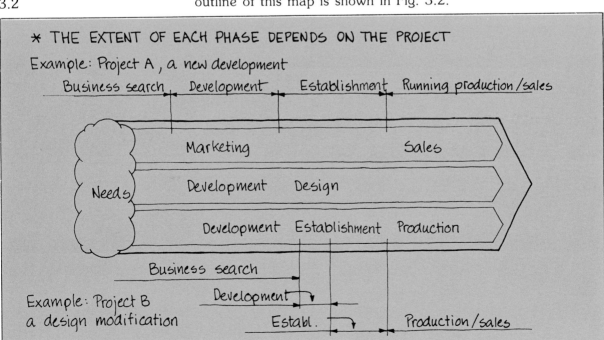

● WE KNOW THE SEQUENCE OF EVENTS IN DESIGN ACTIVITY!

In the areas of work study and design, a general approach for solving problems has long been recognised. (See diagram on the right [3].)

This basic sequence appears at many levels in the design process; for example during formulation of the problem, searching for principles, determination of form and structure, choice of materials, and so on.

The so-called Theory of Properties states that a machine or mechanical product is defined by its basic properties: the structure of the whole product, and the form, materials, dimensions, surface and tolerances of the individual elements.[4]

On the basis of the Theory of Properties, the design process for a product can be illustrated as shown below, in the so-called Product Synthesis [5,6].

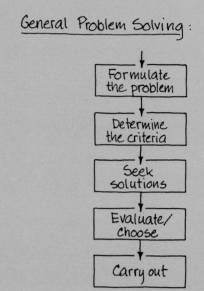

General Problem Solving:

- Formulate the problem
- Determine the criteria
- Seek solutions
- Evaluate/choose
- Carry out

Theory of properties:

Structure

Form
Material
Dimensions
Surface
Tolerances

⇒

All properties:
Function
Price
Lifetime
Design
Reliability
Weight
Strength

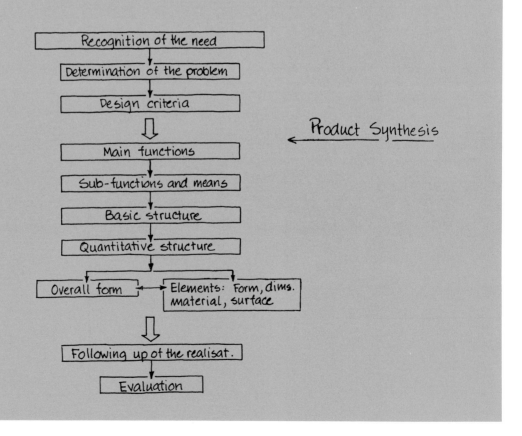

Product Synthesis

- Recognition of the need
- Determination of the problem
- Design criteria
- Main functions
- Sub-functions and means
- Basic structure
- Quantitative structure
- Overall form ↔ Elements: Form, dims. material, surface
- Following up of the realisat.
- Evaluation

24

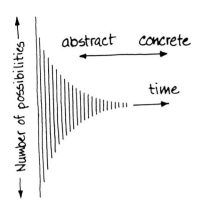

Searching for ideas and choosing between alternative possibilities is at the same time an abandoning of alternatives. That is to say, as we proceed through the development programme and consider, generate solutions or make decisions, then we converge (with fewer and fewer possibilities to 'kick around') towards a particular product which involves all the many choices. At the start our choice revolves around a number of primary, abstract matters such as the type of need and the sort of product that will meet that need; later on, we determine principles and we determine the product's use and structure, the form of the components and so on, down to the smallest details. But throughout this process we choose among alternative possibilities and opt for one of many solutions. If we slice across a development programme, we find we have used up a number of possibilities corresponding to the more important tasks, and there are a series of less important tasks in front of us, for which we have not yet reduced our range of possibilities.

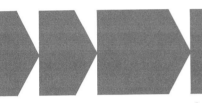

Product development consists of market research, development, the establishment of production and sales, and ongoing production and sales

In the first of these phases, the perceived need is transformed into a form of business which is attractive to the company; recognition of business potential involves the presence of desirable qualities in the market, in the product or in its production, and these desirable qualities must be created in the development phase. The result of the development phase is to clarify which technologies are applicable, so that in the establishment phase the details of the product can be defined in an interplay with the development and establishment of a production system. At the same time, the sales apparatus is established, so that the final phase, ongoing production and sales, can be started. The extent of these phases depends heavily on the nature of the project, as shown in Fig. 3.2.

Design activity determines how the phases are divided up

The study of design over the last few years has provided us with a clear picture of the sequence of events in the activity of design. Roughly speaking, the task of design consists of:

- Detailed analysis of the need.
- Determination of the type of product (for example, deciding that the problem can be solved using an electronic product).
- Design in principle.
- Elaboration of the product (and proving its function).
- Maturation of the product (fixing details to suit volume of sales and production methods).
- Adaptation of the product (adjustment to those problems which turn up when it is launched onto the market, together with any subsequent adaptation).

Quite specific marketing and production activities must be carried out in time with, and in collaboration with, the work of designing the product. There are five distinct phases:

- The investigation of need: Here the marketing task is to establish the basic need, i.e. to identify the need and to decide on the transformation which will meet that need – for example, the need for information on the functioning of the heart, satisfied by the use of electrodes and amplifiers which can transform the electrical activity of the heart into visible heart signals. This also determines the product type, and certain considerations with respect to process type can be put into effect (based on knowledge of probable electrode and amplifier technology).

- Product principle: Basic design work assumes that the way in which the product will be used has been determined, that the man/machine interaction has been made clear, and that the user has been identified. Definition of the principles behind the product and its composition then make it possible to determine the type of production technique which is to be used.

- Product elaboration: Even if the primary goal here is to define more details of the product, and to demonstrate that the product works, it is important that the details of the product are chosen to suit a technology which matches the expected market volume or the selected strategy for volume adaptation. The design work gives a reasonably certain calculation of costs, and the principles of the production system can be developed or determined.

- Production preparation: The goal of this phase is to demonstrate that the product can be produced. For this reason, product maturation should go hand in hand with the definition of the processes and the way in which the product will be assembled. Once the production system has been established, the necessary demonstration can be performed by means of a preliminary production run. Another aim of this phase is to set up the sales system, and to plan how sales and production will adapt dynamically to one another to ensure the best possible launch onto the market.

- Realisation: The content of this phase is ongoing production and sales.

All these considerations can be collected together in the model for 'Integrated Product Development', with the named phases and activities shown in Fig. 3.3.

The integrated product development model basically gives the 'longest path' from a perceived need to sale of the product. In the case of a project with more well-known solutions – for example, where the principles are established and certain details are given – some of the phases will be very much condensed, but the basic sequence of events is the same for all types of project.

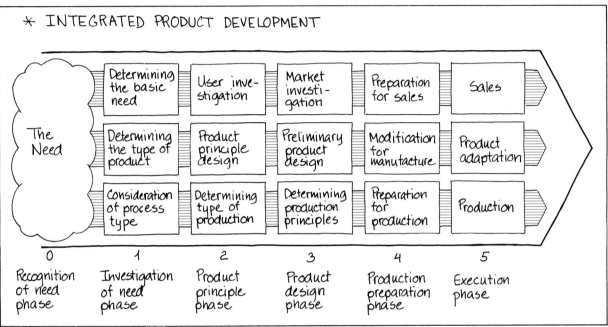

* INTEGRATED PRODUCT DEVELOPMENT

The Need	Determining the basic need	User investigation	Market investigation	Preparation for sales	Sales
	Determining the type of product	Product principle design	Preliminary product design	Modification for manufacture	Product adaptation
	Consideration of process type	Determining type of production	Determining production principles	Preparation for production	Production

0	1	2	3	4	5
Recognition of need phase	Investigation of need phase	Product principle phase	Product design phase	Production preparation phase	Execution phase

3.3

The transitions from one phase to another are important indicators on the way towards good business. The output from the first phase is a perceived need, defined by a product type and process type, and of rather a basic nature – i.e. 'someone' has the need, but who will buy and at what price?

The output from the second phase is a clarification of the product's use and its general principles, which elucidates which type of production can realise these principles. It is important to consider the way in which the product is used and the user/product interaction as something which can be developed. On the basis of the principles of the product, its relations to competing products can be determined.

The output from the third phase is a demonstration that the product works. The size of the market has been estimated, and the feasibility of production processes has been determined. Costs have also been determined to within a reasonable margin of certainty.

The output from the fourth phase is a demonstration that the product can be produced with the desired quality. Production and sales systems have been set up, and prices can be fixed. The output from the final phase is the decisive proof: the product can be sold.

27

• WHAT ABOUT QUALITY, MONEY, PURCHASING, DESIGN STYLE...?

The intention of quality control (QC) is to ensure that the quality of the produced products is uniform and in accordance with their specifications, that the quality of the design corresponds to the aims in mind during development of the products, and that the quality offered to the user corresponds to the user's needs and quality criteria. The first two points reflect the internal efficiency of the company. The last point, which involves interpretation of the customer's desires with respect to quality, and commercial exploitation of these desires, is an aspect of the external efficiency of the company. The quality control function does not have any independent place in the model for Integrated Product Development, but must be thought of as looking after quality-related aspects of everything which takes place within the three functional areas.

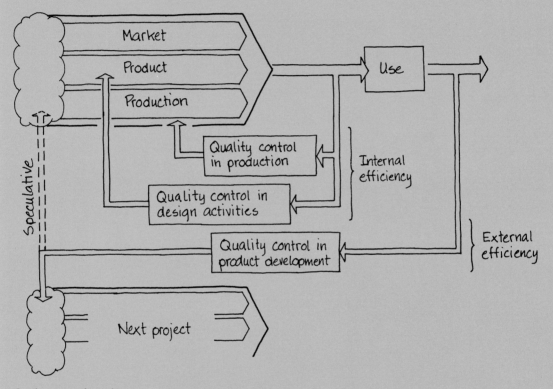

Styling is also design
Industrial design or styling is a particular aspect of the definition of a product, with special emphasis on aesthetics, appearance and man/ machine relations. But what goes on is in principle the same as in the other design activities, just with greater professionalism and more attention to detail in the areas mentioned. Thus there is no independent arrow for styling.

Control is many things
We have chosen to let the model include those activities which are related to specification and creation, and to omit the multitude of controls which are applied to product development and the daily running of the company: financial control, control of salaries, production control, materials control, and so on.

Purchasing
The purchasing function appears to be a poor cousin in the model for Integrated Product Development. It is not included because, as already pointed out, it is not a synthesis activity. However, it plays an important role as regards integration, since purchasing is a connecting link to product development and production by subcontractors. This complex problem is dealt with elsewhere [7].

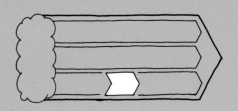

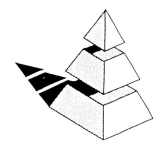

3.2 Integration between project and management

A company is an organism which must continually be monitoring its situation and actively taking steps to maintain or improve its opportunities. The organism is difficult to control, so one must pursue formulated objectives and clearly defined paths toward the target, and also adjust both targets and paths suitably quickly, in order to respond to new situations and circumstances

A company's ability to fulfil its objectives is founded in product development

Product development is an expression of the desire to survive as a company in the long term. Development projects can take 6–10 years to bear fruit, so it is vital that the company's objectives, policy and strategy be incorporated into the projects. Following Eekels [1], we use Fig. 3.4 to explain the forms of control which take place within a company which is pursuing long-term objectives. It is a closed loop. Let us start from the developed products which are being sold on the market, so that there is an economic basis for the efforts to develop new products.

The financial results, market conditions and internal conditions within the company can give rise to adjustments in the company's objectives; for example interest in new markets. On this basis, and on the basis of internal and external conditions, it is possible to decide the company's policies – i.e. major decisions about courses of action which it is thought will lead to good results. One might, for example, concentrate on becoming the leader in a certain market sector, or on achieving a name for quality or design, and so on. Patterns of action in the short term can then be determined from particular strategies, i.e. decisions as to how we shall attain the chosen objectives and results.

3.4 (after(1))

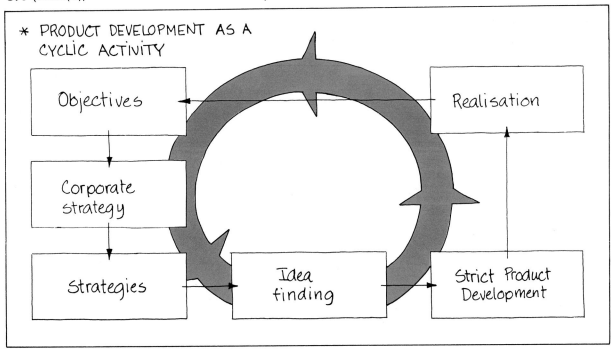

Strategies ought to be developed for many situations within the company and in its relations to the world outside, most importantly for product development as a whole and for its constituent elements, i.e. marketing, design and production strategies.

On this basis, and on the basis of the company's current situation (the need for new business), its willingness to take a risk and its level of ambition, product development can be set going with as more or less radical or 'open' search for business ideas, as shown in Fig. 3.2. When a business opportunity has been identified – i.e. a favourable interaction between a market, a product and a production technique has been found – then the actual development phase is initiated, and this should result in a new product with good earning power on the market. This closes the circle.

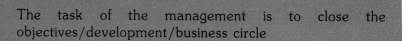

The task of the management is to close the objectives/development/business circle

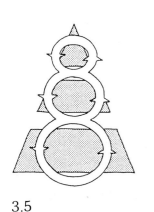

Figure 3.4 does not mean that separate departments or functions should be set up to deal with this task, but rather that there should be a general awareness of it in the company. The small company offers favourable conditions for obtaining an overall view, but in many cases the company does not have enough in reserve for the more long-term considerations (Fig. 3.5A). On the other hand, good results can be obtained by well-timed and highly dynamic adaptation to circumstances. In a large company, there is a greater risk that long-term considerations do not get the attention they deserve, because they are the responsibility of many people. But even where the necessary management task is recognised, or the relevant functions are established, it can easily come to work in a disconnected manner, as shown in (Fig. 3.5B). The reason for this may be 'concrete ceilings' in the organisation, which manifest themselves as a lack of connection between strategies and project objectives.

3.5

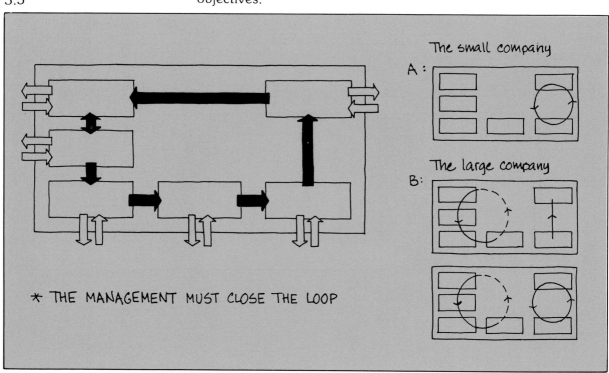

★ THE MANAGEMENT MUST CLOSE THE LOOP

The small company

A:

The large company

B:

3.6

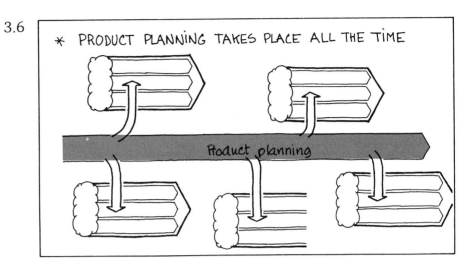

Product planning implies consciously looking after the objectives/development/ business control loop. As sketched in Fig. 3.6, this should be an continual activity, setting product development projects in motion against a background of indications from the world outside, as shown in Fig. 3.5. More precisely, the following tasks must be dealt with:

- Formulation of product strategy, i.e. the more specific strategies which apply to a particular type of product.
- Searching for business ideas and screening them, i.e. concentration on certain characteristics with respect to market, product and production, perhaps also with determination of sales outlets, product principles, production technique, etc.

3.7

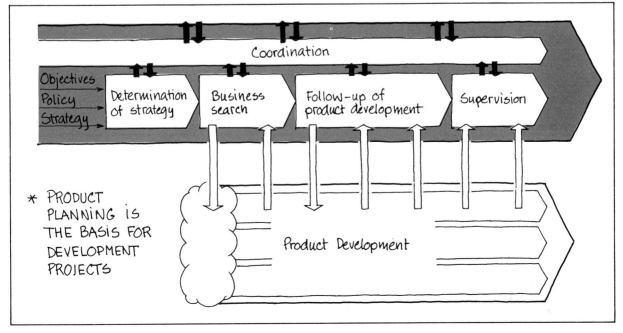

- Initiation and following up of product development projects.
- Top-level control of resources.
- Coordination of the current project in relation to other development projects and other activities within the company that require resources.
- Monitoring the results of product development.

Figure 3.7 shows how the activity of product planning forms a part of and works together with integrated product development.

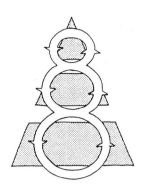

Product development and innovation cannot be confined to a part of a company, but must be an integrated part of the company's efforts to fulfil its objectives. In a company with a complex structure, these activities give rise to several sequences of management activities, for example at a strategic, a tactical and an operational level. As shown in the inset, these activities become part of the cement which must bind the various layers within the company together as a whole.

3.3 Integration with other development activities

In order to adapt to the world around it and to steer actively toward its objectives, a company must not simply develop products, it must also be innovative with respect to technology, methods, strategy, organisation, and so on. Development activities in these areas must act together with and support product development.

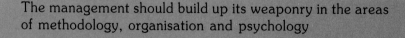

Product development demands professionalism in marketing, in technique and in leadership

Professionalism can be regarded partly as attitudes, skills and knowledge, and partly as the extent to which ready results are available for product development projects. Both areas require innovative initiatives if above average results are to be obtained.

Demands for dynamic and optimal use of resources require interaction between projects of various types, if product development objectives are to be reached (Fig. 3.8).

The management should build up its weaponry in the areas of methodology, organisation and psychology

Innovation requires motivation and capacity. The company should not only be geared up to wanting to develop itself, and have the necessary resources to do so, it must also be geared up with respect to:

- Methodology, that is to say courses of action, procedures, methods, tools, communication techniques, illustrations, forms, etc., which are crucial for the creation of results from product development.

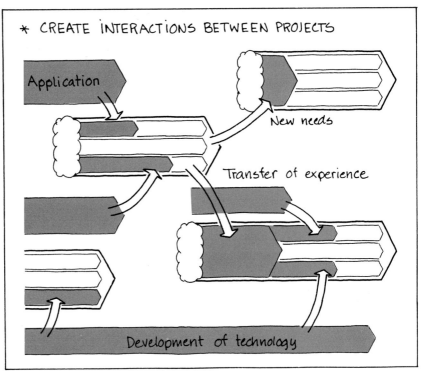

3.8

- Organisation, that is to say the pattern of roles played, administration, the assignment of tasks, responsibility, interaction, grouping, etc., which govern the company and the project group carrying out the project, and all those who are affected in some way by the activity of product development.
- Culture, that is to say creativity, motivation, psychology, behaviour, standard of values, traditions, etc., which to a considerable extent are crucial for the efficiency and quality of product development.

It has been the traditional view for a long time that one should concentrate solely on the organisation as the manipulable factor in adapting a company to changes in internal and external conditions. The need for 'good procedures' and our awareness of the significance of the internal culture of a company have received scant attention.

In this book we concentrate on courses of action and patterns of conduct, derived from the nature of product development, in order to ensure that product development can take place on its own terms rather than on those of the organisation.

• A SLICE OF SALAMI . . .

In a company, many types of development and renewal processes take place. This is illustrated here by letting each such activity have its own timescale in relation to the timescale of a particular development project ('Project 16').

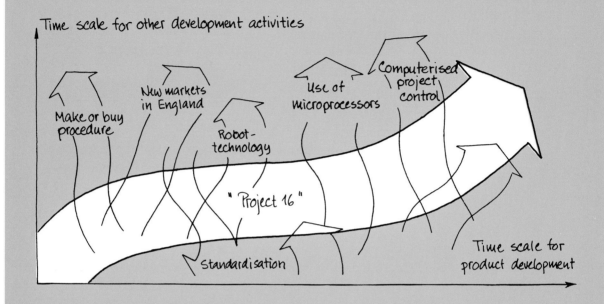

If we took a sectional view through the length of 'Project 16', we should be able to see tracks, results or interactions with the other development activities within the company. If these activities are not organised to the advantage of product development, they are probably doing harm. Integration between development activities appears like the fatty bits which we can see if we slice into a salami. Whose job is it to establish this type of cross-integration?

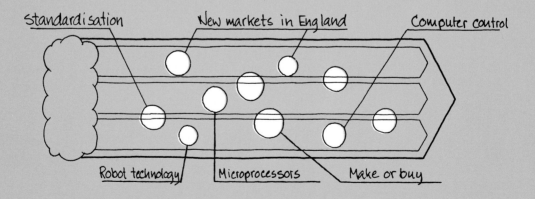

3.4 Understanding the IPD model

The 'Integrated Product Development' model (Fig. 3.3) is presented here as a general model which can describe any product development situation. Before we take a closer look as its validity, we shall explain how the model is to be understood. We can get a good general idea if we make clear what can be considered fixed and what is variable.

If different development projects are compared, we see that the intensity of activity in the individual phases varies markedly: in some projects a considerable development effort is required, in others the emphasis is placed on the design of details or the development of the production system or the creation of new market outlets. But irrespective of the course taken by the project, the following should apply:

> **Every product development project should take the need as its starting point so as to ensure a secure base**

This principle is to be understood in the following way: even if a project has its origins in extremely concrete aspects of an existing product, and for example has as its aim the reduction of production costs, then the previous phases should nevertheless be run through – at least as regards factors that have changed (see Fig. 3.9). The reason for this is that existing business is normally based on considerations and assumptions that may be many years old, and

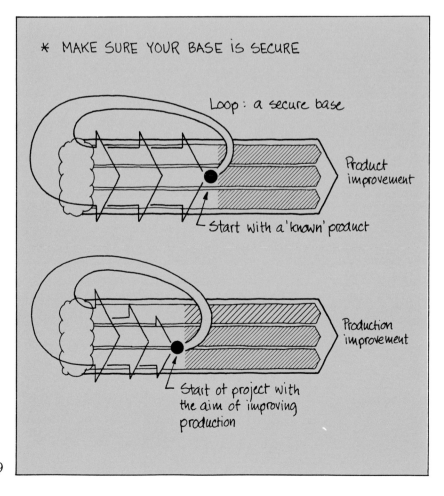

3.9

● A MODEL MADE FROM SIMPLE MATERIALS

The adaptation of the Integrated Product Development model to a particular project can be illustrated with the help of an imaginary visual aid, a 'model board' with the IPD model drawn on it, some elastic, some strips of wood, some nails and a hammer.

The pieces of elastic represent 'what happens' with respect to the market, product and production, and thus have a complete sequence of activities traced onto them. Activities can be pulled out or pushed together, but they cannot be jumped over.

There are three fixed points in the course of development, namely the starting point for development, investment and release of the product onto the market. The three crosswise strips of wood represent these fixed points, and we can start by nailing the last of them on at the start of the last phase in the IPD model. Where other two are to go will depend on the particular situation considered.

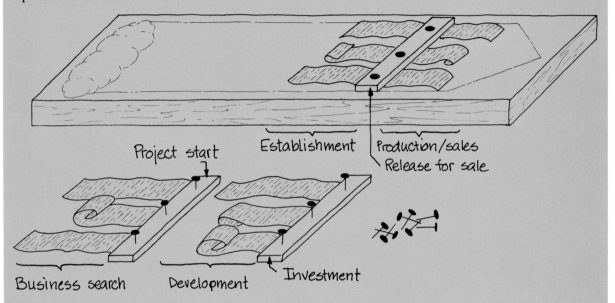

Now we use the model board to describe 'Project 16', which has the properties that the principles for the product are fixed, and that we do not wish to expend a lot of resources on a comprehensive search for business ideas. The production system already exists, and just needs to be modified.

The start-of-project strip is put on at the start of the detailed design phase, and the investment strip in the establishment phase, as we envisage that investment will be moderate, and will be made at a late stage. We then stretch the elastic in those phases where we need a special effort. But we do not cut anything out. And we don't start right in the middle, but secure our base by checking which conditions within the market, product and production may have changed.

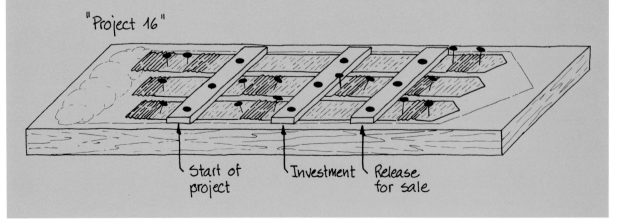

which can have changed considerably in the meantime. The new project may therefore have completely new conditions; for example, with respect to market demands or production technique. A secure base must therefore be ensured. In other words, a new project should as best it can make use of newly recognised market, product and production conditions.

The course followed by different development projects should also be considered unvarying with respect to the results which are attained on the way:

> **At all times the results which are related to market, product and production must be keeping pace with one another**

The reason for this is firstly that development efforts should not be devoted, for example, to defining the product, if the market and production have not been brought on to a similar level of development. Secondly, development work should take place at a pace that is dictated by the regularity of the relationships between the different areas. We return to this subject in Section 6.3.

Keeping pace

The ideal of simultaneous progress in the three areas cannot be attained, but we should strive towards it. Three events (keypoints) in a development project should, at any rate, be given special attention:

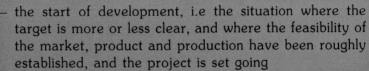

> - the start of development, i.e the situation where the target is more or less clear, and where the feasibility of the market, product and production have been roughly established, and the project is set going
> - investment, i.e. the situation where decisions on investment in production equipment are made
> - the market launch, i.e the situation where the production is a reality and the products are to be released onto the market

A keypoint is a checkpoint at which it is of extreme importance to measure the results of the project and the course which it has followed, and perhaps even to stop the project or give it a shot in the arm. This is the reason for choosing exactly these three situations – they are decisive for the project. This is discussed in more detail in Section 6.6.

Three fixed points

The point at which the product is launched can directly be positioned on the IPD model, while the other two fixed points may be placed more freely depending on the specific situation in hand. In the case of a known product and an established market and production, the development phase can be a very short part of the 'product elaboration' phase. If on the other hand it is a question of searching for new principles, then the development phase will include the whole of the 'product principles' phase.

Of fundamental importance are those features of a new project which are different from earlier projects, and from other projects which possess other market, product or production features. One ought therefore to make a conscious effort to recognise the particular conditions which apply to the specific project one is about to start on, and to create the project on its own terms. We shall return to this in Section 6.6.

What are the variable factors in a development project? What distinguishes projects carried out in different types of company? Later in this book we shall look at many of these factors in detail; at this stage, the following variables in product development projects are dealt with:

- The area of renewal (market/product/production).
- The degree of renewal.
- The frequency of renewal.
- In-house/outside development.
- The relationship between development costs and sales costs.
- Production to order/mass production.
- The extent to which the project is predetermined.

The area of renewal

The desired renewal to be brought about by a development project may of course lie in one or more of the three areas of market, product or production. In this book we shall principally consider situations where renewal involves changes in the product itself.

The degree of renewal

Product development can be divided into four main phases (Fig. 3.2), namely, the search for business, development, the establishment of production and sales, and ongoing production and sales. These four basic phases can only partially be associated with the division into phases shown in the model of Fig. 3.3, as the search for business can occur at many different levels, right from a search for a radically new product, completely new markets and/or completely new production technologies, to a search for product changes or better adapted production equipment as a basis for improving an existing business opportunity.

Traditionally, the degree of innovation has been described in a two-dimensional diagram (Fig. 3.10), where product technology and production technology are combined. As regards the product, it is usual to distinguish three degrees of renewal, namely new design, adaptation of a design, and variation of a design. As regards production, one can roughly speaking use a division on three levels: new production processes, new production equipment (using

3.10

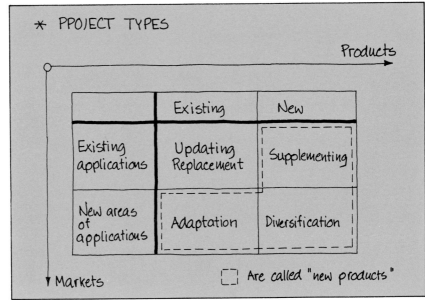

* PPOJECT TYPES

	Existing	New
Existing applications	Updating Replacement	Supplementing
New areas of applications	Adaptation	Diversification

Products

Markets

☐ Are called "new products"

comparatively well-known process machines), and the adaptation of existing equipment to the task in hand. With respect to the market, one can distinguish between completely new markets (new sales outlets, establishment of a sales organisation), the broadening of existing markets by building up the sales organisation, or the use of known markets, i.e use of an existing sales organisation.

The degree of renewal in a company is of course a relative concept: in one company, the changeover to new packaging may be considered significant renewal, whereas renewal in another company might be based on looking for an unnoticed need. The positioning of the cloud in the IPD model will therefore mean different things, depending on the reader's background and present situation. Our intention is that the cloud should represent known or unknown needs and that the path from these needs to a concrete solution should go via decisions about the product's type, principles, design and adaptation to production. It is not really a question of the cloud being placed in different places, merely that we should accept larger or smaller blank and unused areas on the map (Fig. 3.11).

3.11

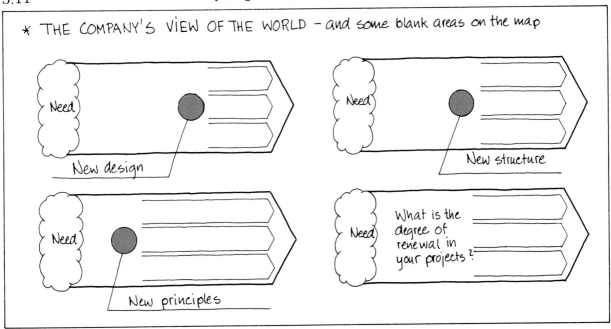

* THE COMPANY'S VIEW OF THE WORLD – and some blank areas on the map

Need
New design

Need
New structure

Need
New principles

Need
What is the degree of renewal in your projects?

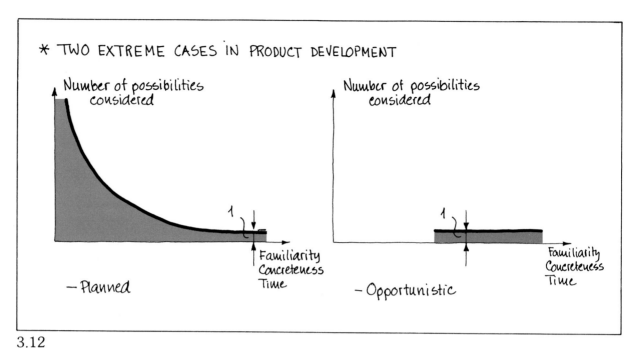

3.12

The frequency of renewal

Renewal activity in companies covers the whole spectrum from product development being a rare event (or one which never occurs) to it being a frequent occurrence; and from small steps to large, high-risk projects. There is an important dividing line here between planned and opportunistic product development (Fig. 3.12). The first type is characterised by a broad search among various possibilities, gradual reduction in their number, gradual adaptation and goal-setting and finally the execution of a specific course of development during which many development projects, introduced to support the project finally undertaken, are usually left incomplete or abandoned. The other type or extreme is the company which, on the basis of impressions or intuition, launches into a definite project and carries the project through whatever the cost. Between these two extremes we find a whole range of intermediate types.

An important characteristic is the taking of risks versus safety at the launching of a product. Some companies choose to carry out

3.13 (after(16))

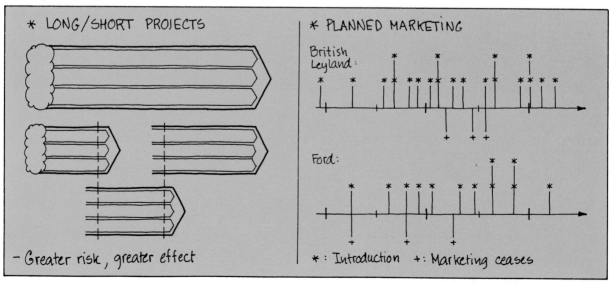

40

protracted projects, resulting in very 'finished' products, before marketing takes place, while other companies in the same period of time market several 'unfinished' products, thus taking a greater risk, but with better adaptation to the market as a result (Fig. 3.13).

In-house/outside development

Larger or smaller parts of a product development project can be carried out as out-of-house development, so that areas of the IPD model are associated with activities outside the organisation. The extent of such an area and its relation to the remainder of the IPD model tells us a good deal about the problems that arise in the two regimes. In Fig. 3.14 we see some important types of development work which contain elements of out-of-house activity. We shall not discuss these various possibilities in more detail, but merely point out that many different forms exist, each with its particular problems which arise because we sever certain natural relationships:

- Products are developed which are to be realised by a subcontractor, i.e. without exact knowledge of his production conditions and development.

3.14

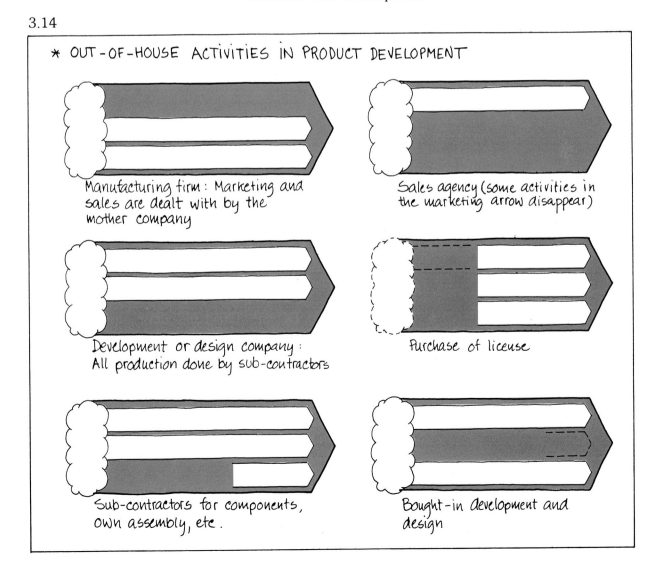

* OUT-OF-HOUSE ACTIVITIES IN PRODUCT DEVELOPMENT

Manufacturing firm: Marketing and sales are dealt with by the mother company

Sales agency (some activities in the marketing arrow disappear)

Development or design company: All production done by sub-contractors

Purchase of license

Sub-contractors for components, own assembly, etc.

Bought-in development and design

- Product development is carried out in, let us say a development centre, without being precisely related to the company's strategy and tradition, or to the market and production.
- Components are required which must suit a current product, and must be developed and produced by another company.
- A license is bought, i.e. a relatively complete line of business, which has not been adapted in advance to the company's conditions, and which is to be taken over by people who were not involved in creating it.

So there are a number of tasks when we cut across the product development model both horizontally and vertically. The cuts may be beneficial in that one becomes aware of important relationships and tasks, which must be carried out, and they may be adverse in that they distort the nature and coherence of product development.

The relationship between development costs and sales price (DC/SP)

The DC/SP ratio tells us how many units of the developed product need to be sold before development costs are nominally recouped. In what follows, a 'reasonable' ratio between sales price and costs is assumed. DC/SP can vary from 0.1 to 1,000,000. Roughly speaking, this range can be divided into three categories [8]:

- Very low DC/SP (<20): The products are expensive, for example diesel engines for ships, cement factories, offshore rigs, bridges, water purification plants, large production plants, large production machines. Development takes place in small steps, the company is based on the accumulation of experience, production has a low priority, analysis and knowledge of parameters are crucial.

- Very high DC/SP (> 1,000,000): The products are very cheap in relation to development costs, and as little as possible is devoted to new development. Marketing is vital for business (consumer products such as toothbrushes, toothpaste, bicycles, foodstuffs). The company is market-oriented and puts its effort into trading, often with sub-contractors with a similarly high DC/SP ratio (beer cans, bicycle parts).

- Medium range DC/SP (20–1,000,000): In this range, development costs are significant in determining the unit price and the market is typically unsettled: short life, large supply, great competition. It is important here that business opportunities are maximised, i.e that with the overall picture in mind efforts are concentrated onto marketing, product and production, and on development in all three of these areas. In this field we find a series of products from mechanical industry: engines, valves, pumps, refrigerators, etc.

Integrated product development can be applied to greatest effect in the medium range category above. Consideration of the DC/SP ratio does not, however, bring us nearer to being able to point to specific factors or general rules which are dependent on DC/SP.

Production to order/mass production

The property of production to order/mass production is of course closely related to the DC/SP ratio, but includes a time factor which we will look at in more detail. In a later chapter (Chapter 8), we shall see how there is an important relationship between those phases in which the product is designed, and the dynamic properties of the production, including its adaptation to production to order or mass production.

Production to order/mass production can be a strictly localised question of production technique, where for example items are being produced for storage between the handling of larger special orders. But things become more complex when there are special sales or development activities associated with each order or production series, perhaps with special delivery conditions and variation or adaptation of the product. In the IPD model, this can be expressed as a cut between product development and production to order/mass production (see Fig. 3.15). In this figure, a cutting point has been introduced at the development stage on which all orders are based. After the cutting point, there can be a greater or lesser amount of development work on the basis of the common foundation laid before the cutting point.

As emphasised earlier, it is important that projects are carried out with market, product and production held in step with one another. This also applies in the case of the 'cut' made for special order production. At the cutting point, special demands are to be made with respect to the creation of variations, adaptation of the product, and flexibility and dynamics in the capacity and output of production.

3.15

* PRODUCTION TO ORDER accompanied by ...

.. no   .. small   .. large development efforts

Production to order:

Order 1

Order 4

Order 3

Order 5

Order 2

Cutting point:
Market/product/production have been fixed.

Order:
Customer/product variant and production run are fixed

Activity up to the cutting point and the 'availability' of the cutting point are important for companies with low DC/SP ratio (bridges, mechanical plant, large production machines, such as systems for automatic assembly or the fabrication of concrete pipes, and large, specialised food processing machinery) and those with higher DC/SP ratios (agricultural machines, cooling plant, reproduction cameras, medical measuring equipment, etc.).

The extent to which the project is predetermined

By this is meant the extent to which a development project is based on known and accepted factors, and therefore will tend merely to be a form of planning. We can also make a cut for predeterminedness, and thus isolate the preliminary work which is either carried out once and for all within the company for the product concerned, or which one has no intention of tampering with (for example, because one would not be able to do so, because it demands new development). But apart from this, parts of the production (e.g. processes, production equipment, subcontractors, surface treatment, packaging), and of the marketing (known markets, known sales outlets, known clientele, already determined price policy, etc.) may be predetermined.

The degree of predeterminedness has great significance for the handling of project risk and for project planning. We deal with this question in a number of places in this book; for example, as regards identification of the project situation (Section 6.6), project planning (Section 6.8) and dealing with risks (Section 7.2).

3.5 Validity of the IPD model

Companies develop through a number of characteristic phases (see Fig. 3.16), and each phase has, of course, its own characteristic problems. Problems of lack of integration, for which the message and methods of this book provide a solution, are present in several phases in that there are many aspects to 'disintegration' within a company, as we saw in Chapter 1. We shall concentrate on the large company that is heavily divided up according to function; however, the smaller company will derive considerable benefit from seeing the strengths and weaknesses involved in being integrated.

Integrated product development is addressed to companies that are characterised by the prevailing project philosophy in Western countries:

- Product development is carried out by engineers who have been trained in specialised disciplines and appointed to technically specialised positions.
- These engineers seek to safeguard their future by increasing specialisation, and by development of the area they are assigned to look after, in preference to making a contribution to product development and the company as a whole.
- They work predominantly in groups, in order to take advantage of the benefits to be had from the creative interaction between several people, but more importantly because no single person has the necessary knowledge and experience.
- Development is for the most part governed by procedures.

3.16

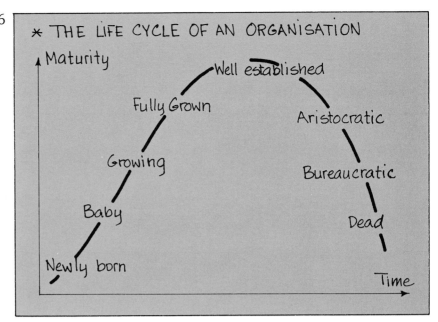

* THE LIFE CYCLE OF AN ORGANISATION

- Maturity
- Well established
- Fully Grown
- Aristocratic
- Growing
- Bureaucratic
- Baby
- Dead
- Newly born
- Time

- Employees are involved in several different types of activity at one time, with high and low priorities, with short and long timescales, most often with very few people involved, and with a low priority assigned to development activity.

Integrated product development is intended for use by companies with many levels and thus with problems in getting the top-level objectives and strategies out to the employees, and in getting everybody motivated. But the management of smaller companies can also obtain advantage from the perceptions which are associated with integration, because the special strengths of small companies are emphasised.

The authors' knowledge is related to companies in mechanical, electromechanical and electronic industries. The ideas considered in this book are not limited to particular types of product, production methods or markets. Different indicators may be brought into play for describing types of company and types of development activity. We shall now take a closer look at this, and comment on the organisation and validity of the IPD model in relation to it.

It is assumed that the companies which we are addressing have the following structure in relation to projects which they engage in:

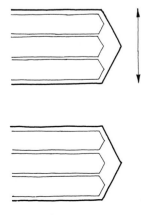

- Crosswise coherent, i.e. with market-related sales, product development/ design of a more or less original nature, and with some production activity (in-house or using subcontractors), but also with significant purchasing. Thus all three areas of the IPD model are present.
- Lengthwise coherent, i.e. with a more or less distinct division of project activity into phases, under the control of a more or less conscious product planning activity.
- With a significant element of uncertainty and/or risk as regards technology and/or market conditions. In this respect, development projects differ significantly from mere projecting, where the result is so well known that almost everything can be allocated deadlines, without there being any risk that certain sub-tasks cannot be achieved in time (Fig. 3.17).

● A MODEL WITH EMPHASIS ON FINANCIAL AND ORGANISATIONAL ASPECTS

As mentioned in the Preface, Fredy Olsson of the University of Lund [13] was one of the pioneers of integrated product development. The latest edition of Fredy Olsson's model of product development, now in use within the Swedish Mechanical Engineering Association (Mekanförbundet), looks like this:

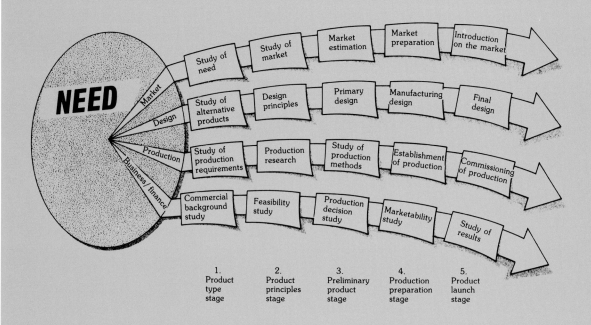

The principal difference from the IPD model is the fourth arrow, which deals with the creation of a financial and administrative environment. Thus the model is well-suited for use in the setting up of new companies or subsidiaries, where finance, borrowing and the establishment of new business are in focus. The fourth arrow is partly to do with forecasting business, and partly to do with its creation. The aim of each phase along the arrow is accurate prediction of the consequences of what is to be created, and the establishment of the financial prerequisites for this.

In our Integrated Product Development model, we have chosen to omit the fourth arrow, as we are addressing ourselves primarily to companies with established product development, and therefore normally in a well-established financial situation.

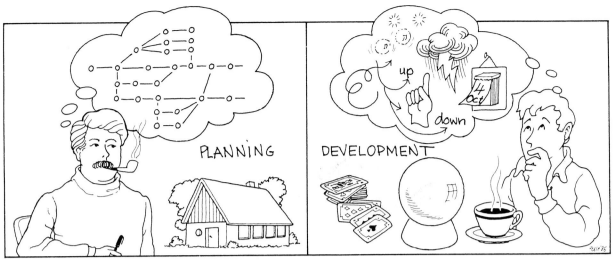

3.17

The objectives and situation of the company are also important for the usefulness and significance that integrated product development will have. More often than not, the current situation is of little interest; of much greater importance is the question of where the company should have got to, and how it reacts in response to external conditions. Here are some characteristic external conditions and the types of company associated with them [9,10].

- Companies with a narrow market and product range, such as cars, diesel motors for ships, government protected companies in the energy and telecommunications sectors (cable and switching equipment, telephones), the bicycle industry, and so on. This type of company is highly specialised, geared up for this type of activity and without any desire to seek new pastures. They seldom need (or seldom recognise the need in time) to introduce major changes in technology or organisation. Instead, they optimise their activities. They have a tendency to institutionalise their development activities, following a fixed track, and as a matter of routine.

- Companies which grasp every opportunity to pursue marketing aims, and who are aggressive toward competitors. Their development activity is characterised by a high degree of renewal and rapid development. Their production system is characterised by rapid establishment, based on the use of universal machines and standard workplaces. Because of the considerable requirements for a dynamic and flexible response, this type of company rarely has internal efficiency. (Apparatus, such as measuring equipment, imported leisure electronics, auxiliary equipment such as high-pressure cleaning equipment.)

- Companies with a high degree of individualisation in their products, to suit customer requirements, with flexible (e.g. modular) products and a flexible production system, achieved for example by using universal machines and a relatively high degree of mechanisation or automatisation. Here it is a question of special orders and short production times. (Pumps, agricultural machinery, ventilation equipment.)

47

'The story is told, that once upon a time in Prague seven learned professors were led in blindfold to an elephant, with a view to their describing the elephant afterwards. They came to many astonishing conclusions. It was only the elephant's sweet disposition that prevented them from coming to harm in this experiment.'

In the same way, people come to many different results if they are asked to describe the activity of product development, depending on their previous experience, their professional points of view, who they choose to consider within the company, and their reasons for making the description.

Many points of view can be used:

- The organisational
- The creative
- The managerial
- The activity-related
- The strategic

None of these points of view on its own gives a fair picture of the activity of product development. An organisation on its own does not give results, creativity is helpless without professionalism, management and leadership require something to manage, activity is pointless without a target, management and organisation, and strategic considerations require activities in order to be carried out in practice.

In this book we take as our starting point the activity of product development, and attempt, on the basis of the nature of this activity and the circumstances prevailing within the company, to develop guidelines for improving its efficiency.

- Companies who concentrate on a segment of the market with one or more desirable attributes, such as design, service, durability or fashion. In development and production, most attention is payed to these attributes, so quality control is a dominant feature. (Furniture, control equipment, Danish leisure electronics.)

In this spectrum of company situations we find many variations, depending on the company's strategic objectives, tradition, existing know-how and equipment, management and employee attitudes, how the company perceives its own situation, its strategy for change, developments in the market, technological developments in the relevant area, and so on.

We have elected to keep the organisational aspects of product development in the background, partly because organisation theory is not our subject, and partly because many forms of organisation can be 'used' to carry out the same sort of development. We believe that the patterns of behaviour and the attitudes of those who are to carry out development tasks are of much greater importance than organisatorial factors. We will, however, just mention a series of traditional factors, such as the composition of the project group, its position in the organisation and its leadership, together with the identity and position in the organisation of the project leader.

At this stage there is some point in stressing what integrated product development is <u>not</u>, and where the IPD model is no longer valid:

- IPD is <u>not</u> a pattern of <u>behaviour</u> and <u>communication</u> which can put right organisational shortcomings, but a pattern of activity in which simultaneity of events related to the market, the product and its production has its origins in the nature of the product and the nature of the development activity.
- IPD is <u>not</u> a <u>project plan</u>, but a basic pattern for creating attitudes and knowledge, which can be transformed into a project plan through a series of situation-dependent decisions, as shown in Section 6.3.

In IPD, attention is focused on those activities which are aspects of the <u>profession</u> of product development, those things that have to happen. Thus other aspects must necessarily slip into the background – those aspects that are not unique to this profession, such as organisation, creativity, management/leadership, strategy, design, innovation and analytical methods.

The exposition in this chapter has presented the foundation on which the rest of the book is built. In the succeeding, problem-oriented sections of the book we shall show how the integration concept, realised in different ways, can assist in the solution of a series of problems within the company.

It is our hope, that the reader will recognise his own situation amid the problems which we present, and that our answers therefore will serve both to increase his awareness and to inspire him.

We have chosen to divide our treatment of the material into five subjects:

- Get results from product development: It is the management's task to fix the type and scope of product development, to assign the right people to the job, to get these people to be innovative, to ensure they work effectively, to control their use of resources and to make the most of the desired business.
- Create the project: This task lies on the border between the management and the project, and involves both the management and the project leader. The project must be created on the basis of the strategic objectives, it must be created on the terms dictated both by the situation and the product, and it must be set going in the right direction.
- Assess your chances...: Knowledge of the external conditions, competition and risk factors in product development is the prerequisite for being able to take a chance.
- ...and exploit your opportunities: People say that product development cannot be controlled with 100% certainty, but you can do a lot to keep in training, and each individual development project can be exploited in order to re-orientate the company.
- Integration: Here we shall review the potential of the integration concept and our level of understanding. Even if we have a number of tools at our disposal, we need to recognise that there are still many problem areas in which knowledge and tools are lacking.

Before we get down to these five topics, we shall take a closer look at the role of the management in the course of product development, and in the determination of the type of innovation which is most suited to the company.

Development projects and the management

SUMMARY

In this chapter attention is focussed on the management's problems, their tasks and their possibilities for action within the concept of integrated product development. Most emphasis will be placed on project activity as it is described in the basic model, but in order to establish the responsibilities of the management we have to look at the complete product development cycle and at the project as an organisational unit within the company.

This chapter gives an introduction to the basic and professional activities pertaining to product development, that is to say the basic knowledge of the nature of projects which is necessary if the management is to play the right role.

Not all development projects have to start with a broad search for business ideas. There is a whole spectrum of projects, ranging from products that are 'completely new' to products which only the initiated will be able to distinguish from the previous ones. In the last section a quick survey of possible types of project is made.

4.1 Organisation of a development project

The development of a product is normally carried out as a project because it is a task with certain characteristic properties:

- It runs for a fairly long, but definite, period of time.
- It has its own staffing resources – a project group.
- It is executed within well-defined resource limits.
- It spans the entire organisation – those engaged in it must work in a different manner than that defined by the company's basic organisation.
- It is interdisciplinary – it is so complex and demanding that various departments or specialists have to take part.
- It is characterised by development – innovative products are expected to be created.
- It is normally important for the survival of the company and is therefore controlled by, and dependent on, top management.

Some of these characteristics are very important and should be specially noted.

Development projects require a separate organisation

4.1 (after(11))

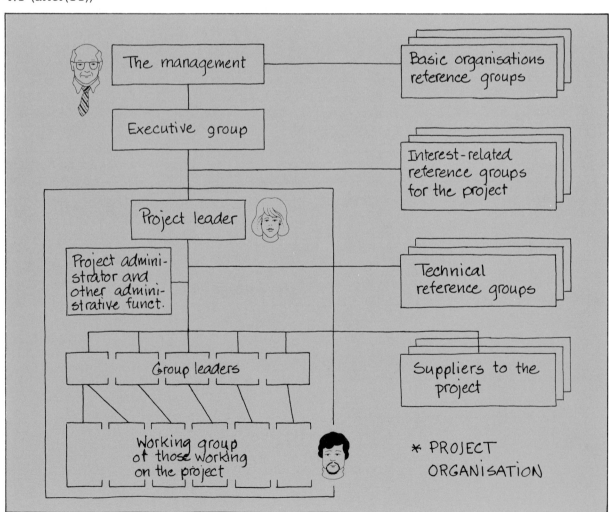

The reason for this is that, during the course of the project, it is necessary to define organisations which are to be responsible for this or that, to select employees, set targets, make plans of action, and so on. In other words, a range of activities which will not fit into the existing pattern of the basic organisation.

As a basic model for project organisation, we shall use that developed by Hans Mikkelsen [11]. Here, a distinction is made between the 'internal' organisation, which is made up of work, project management and executive elements, and an 'external' part of the project organisation, which is made up of a supplier element and a reference element (Fig. 4.1). The nature and purpose of these organisational elements needs some further explanation:

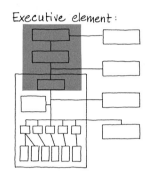

Executive element:

The project's <u>executive</u> element is composed of:

- The executive (or the management) who initiates the project, allocates it the required resources (personnel, money) and 'enforces' targets, policies and strategy.
- The executive group (product committee, project monitoring group) who most often represent the basic organisation which is going to take the product over, and which has the responsibility for the correlation of projects and the setting of their relative priorities. If a group of this type is set up, it looks after the general direction of the project (the project's 'external' management, see Chapter 5).
- The project leader who determines the internal aspects of the project, and presents the results of the project for approval at the agreed decision points (key points, see Section 6.4).

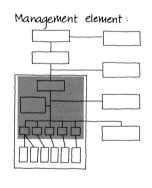

Management element:

The project's <u>management</u> element consists of:

- The project leader who interprets the project objectives, formulates its strategy, methods and support, and who is responsible to the management for the technical result, the keeping of deadlines, the quality and the use of resources.
- The group leaders who are in charge of and responsible for limited parts of the project. This level normally only appears in large projects, and should be dictated by the size of the project, not by its technical composition.
- The project administrator who explains and presents information for use in the management of the project, especially for its control.

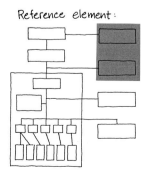

Reference element:

The project's <u>reference</u> element consists of:

- Reference groups or committees from the basic organisation, which are set up if the ideal composition of the executive group would make it too large. These groups express their opinion on proposals and report to the executive group.
- Permanent committees within the company, such as industrial relations committees, safety committees and technology committees.

Supplier element:

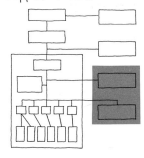

Participant element:

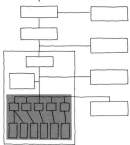

The project's <u>supplier</u> element consists of:

- Technical reference groups, that is to say specialists who perform limited tasks in accordance with the project leader's specifications, or who are involved in order to ensure that particular aspects of the project are looked after. Examples of such aspects are: standards, official approval, patents, packaging, industrial design (this is not how it is supposed to happen, but unfortunately it often does), dimensioning, quality control (also not how it should be, but...), and so on.

And finally, the most important element of the project, the <u>work</u> element, which consists of:

- The members of the project teams (including the group leaders and perhaps also the project leaders) who perform the creative activities within the projects, and who have an individual responsibility for carrying out their personal tasks within the project, and for coordination of their activities with others in the project group. We shall look at this in more detail in Section 5.4. In some cases, project participants will only use part of their time on the project, or will only participate in part of the project.

All this should be thought of as an overall model for the project, a model which only applies to its full extent in the case of complex projects in complex concerns. In small– or medium-sized companies, matters are much more simple, because the same people perform several roles. This increases efficiency and decision-making ability, and makes it easier to take an overall view. But there may, of course, be limitations, if the full range of expertise is not present.

Regardless of the size of the company, it is still true, as pointed out above, that development projects cannot be carried out within the framework of a company's basic organisation (Fig. 4.2).

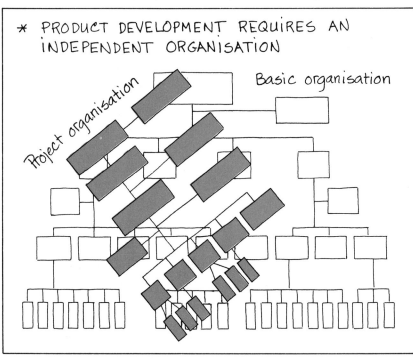

4.2

4.3   4.4

Another feature which is characteristic for product development is the way in which it is carried out – the actual process of 'developing a product':

> **Product development requires the ability to carry out projects**

In addition to being competent in the areas of engineering, commerce, finance, marketing, production and quality assurance, the people who make up the company must also have the kind of competence which ensures professionalism in carrying out projects. Some of the company's employees must be 'generalists' with an overall grasp of the company's activities, insight into the relationships between these activities, and with know-how about how projects are carried out and about the particular strategies, methods, tools, decision models and conditions which are relevant for product development, and its interplay with other activities within the company. But it is not enough that a few persons with this speciality are present; the company must also gear itself up for product development by organising this activity in accordance with its natural characteristics, by carrying project activities out in a professional manner, and by getting the team members to collaborate in this. This orients everything towards development, as shown in Fig. 4.3.

An organisation on its own doesn't ensure results (see Fig. 4.4), so we shall look more closely at the tasks involved in product development and the role of the management in the pattern of behaviour.

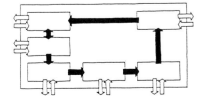

4.2 Role and responsibilities of management

We have pointed out earlier that it is the role and responsibility of the management to close the loop of Fig. 3.5 – that is to say, to create the necessary coherence between company objectives,

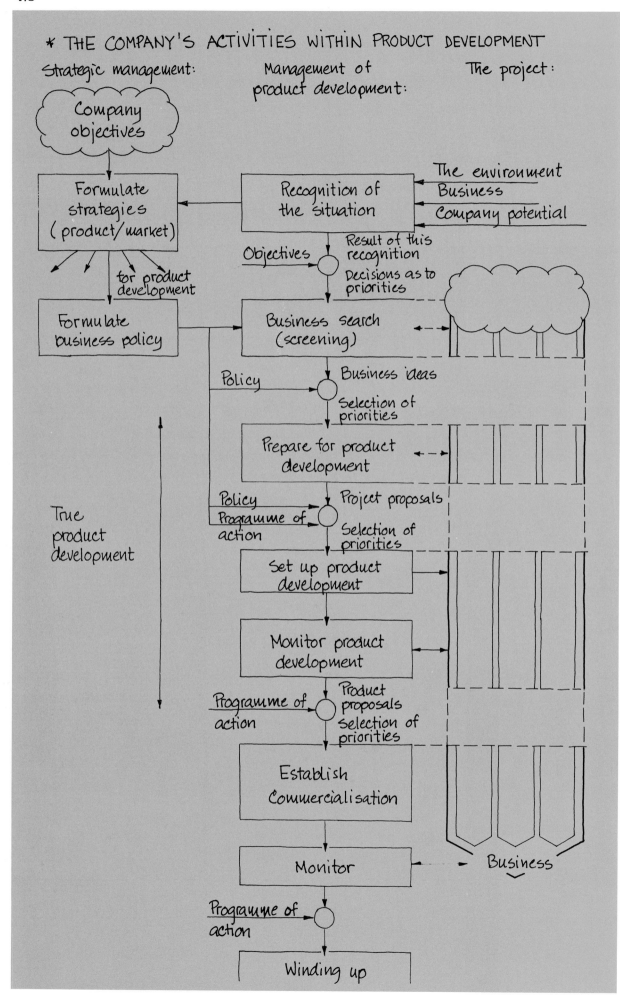

* THE COMPANY'S ACTIVITIES WITHIN PRODUCT DEVELOPMENT

policy, strategy, the search for business, actual product development and marketing.

If we open the activity loop at the place where several products are on the market, then the start of a new development can be changed conditions in the market or within the company (Fig. 4.5). On the basis of results, its current situation or its future prospects, the company must adjust its strategies or formulate new ones in accordance with its company objectives. Efforts to compensate for deviations from the formulated objectives or for exceeding particular limits (for example, if profits are too low, the share of the market is too small, the technology gap to the competition is too large, the production apparatus is too old) leads to activities which, as far as product development is concerned, are initiated on the basis of commercial policy and objectives.

Product development, as it is formulated in the model for 'Integrated Product Development', starts with a business search phase, in which the target is to find potential commercial possibilities consisting of perceived needs, product ideas and/or ideas with respect to production technique. This search can be carried out systematically through the formulation of areas to be searched and the use of idea methods, or the new commercial possibilities may just pop up by themselves or as the more or less random results of participants' considerations or contacts with customers. The result of the search will be some possibilities which can be set going at once, and some which will be kept for use later. Project initiation normally takes place on the basis of a set of priorities based on company policy.

The results of the search for business ideas are normally incomplete with respect to the degree of definition of the three different areas: the market, the product and its production. There may be crucial areas where there is a need for clarification calling for additional effort, for example in the form of research, market investigations or development of processes, which can be carried out as R & D or by the use of technology packages (see Chapter 9). A preparatory phase is therefore set going, with the aim of creating a well-defined project objective and clarifying all factors which would be able to disturb the regular course of the product development project. How far it is necessary to go here in the way of clarification depends entirely on the nature of the company, the level of ambition of the project, and the company's background. Some companies can permit themselves to set ambitious, risky projects going, whereas others must restrict themselves to more run-of-the-mill activities.

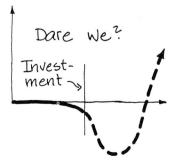

The result of the preparatory phase is concrete project proposals with a reasonably certain guess as to how likely it is that they can be carried through, and what they will cost, but with only a vague sketch of what the commercial result will be and what investment will be required. On the basis of the company's policy and preliminary marketing plans, decisions are made as to whether the project proposal should be started, what priority it should have, or whether it should be kept until later. A good number of proposals have normally to be discarded due to lack of quality or because the opportunity to start them has been missed.

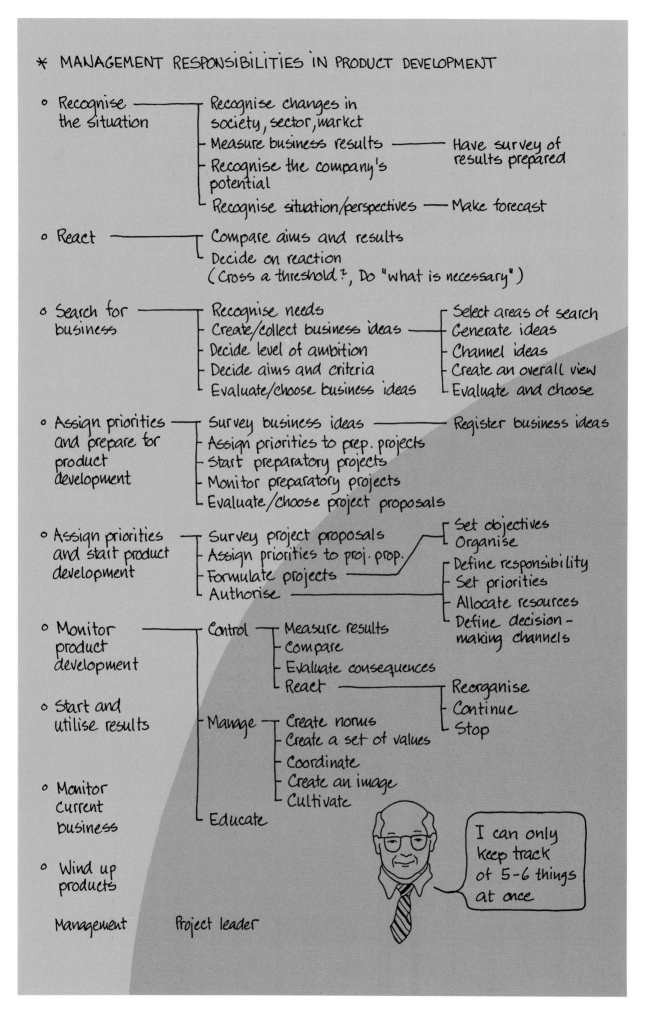

* MANAGEMENT RESPONSIBILITIES IN PRODUCT DEVELOPMENT

o Recognise ———
the situation
  ┌ Recognise changes in
  │ society, sector, market
  ├ Measure business results ——— Have survey of
  │   results prepared
  ├ Recognise the company's
  │ potential
  └ Recognise situation/perspectives —— Make forecast

o React ———
  ┌ Compare aims and results
  └ Decide on reaction
    (Cross a threshold?, Do "what is necessary")

o Search for ———
business
  ┌ Recognise needs   ┌ Select areas of search
  ├ Create/collect business ideas ——   ├ Generate ideas
  ├ Decide level of ambition   ├ Channel ideas
  ├ Decide aims and criteria   ├ Create an overall view
  └ Evaluate/choose business ideas   └ Evaluate and choose

o Assign priorities ———
and prepare for
product
development
  ┌ Survey business ideas ——— Register business ideas
  ├ Assign priorities to prep. projects
  ├ Start preparatory projects
  ├ Monitor preparatory projects
  └ Evaluate/choose project proposals

o Assign priorities ———
and start product
development
  ┌ Survey project proposals   ┌ Set objectives
  ├ Assign priorities to proj. prop. └ Organise
  ├ Formulate projects ——   ┌ Define responsibility
  └ Authorise ———   ├ Set priorities
                                     ├ Allocate resources
                                     └ Define decision-
                                       making channels

o Monitor ———
product
development
  Control ┬ Measure results
          ├ Compare
          ├ Evaluate consequences
          └ React ——— ┌ Reorganise
                       ├ Continue
                       └ Stop

o Start and
utilise results
  Manage ┬ Create norms
         ├ Create a set of values
         ├ Coordinate
         ├ Create an image
         └ Cultivate

o Monitor
current
business
  └ Educate

o Wind up
products

Management   Project leader

I can only
keep track
of 5-6 things
at once

58

The role of the management in the actual product development, which is started on the basis of the project proposal, is to make the decision to start, and to monitor whether the group of current projects together appear, with a reasonable degree of probability, to be leading to the required commercial results. A crucial point here is the decision on whether to invest in production equipment, at which point a careful evaluation of the project must take place. Of equal importance is the decision as to whether to release the product for sale, where the reputation of the company is at stake and where one has to commit oneself in relation to one's customers.

Once development of the product is over, it is the responsibility of the management to monitor what happens to the product after this, to initiate modifications and finally to decide when and how the product is to be discontinued.

The management's roles and responsibilities in connection with product development are multifarious and can be derived from several factors:

- The organisation of projects.
- The activities which have to be carried out.
- The efficiency and quality which are characterise these activities and their result.
- The motivation of the team members and their personal development.

The first two points have been dealt with above, and can therefore act as the starting point for a resumé of what necessarily has to be carried out in connection with product development – that is, the 'functions'. The last two points will be considered in the next chapter.

Figure 4.6 shows the tasks of the management in connection with product development, related to the number of alternatives which are under consideration, and which could thus be the basis for realisable business. On the opposite page, a more detailed version of this functional scheme is given, based on the above considerations. The figure may be thought of as a checklist for those tasks which require decisions during the setting up of a development project.

Figure 4.6 and the opposite page show the basic lines of the professionalism which the management of the company must demonstrate with respect to product development. In the following sections, we shall look at a series of general factors which have to be considered if the result is to be more than just ordinary.

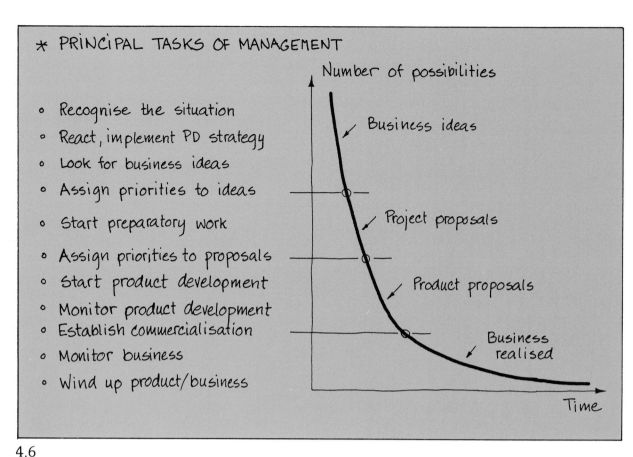

* PRINCIPAL TASKS OF MANAGEMENT

- Recognise the situation
- React, implement PD strategy
- Look for business ideas
- Assign priorities to ideas
- Start preparatory work
- Assign priorities to proposals
- Start product development
- Monitor product development
- Establish commercialisation
- Monitor business
- Wind up product/business

4.6

4.3 Project types and innovation patterns

In Section 3.4 we saw that development projects, despite their differences, have a number of fixed points, and that projects can be characterised by their area of renewal, degree of renewal, frequency of renewal, degree of in-house and external development, the relation between development costs and sales price, the question of production to order versus mass production, and the extent to which the project is predetermined.

In this section we will look at the situation in companies where the degree of innovation is roughly the same in all projects and where there is therefore no need to go back a long way in the development sequence. To put it another way: What does product development look like if we limit ourselves to design modifications?

But before we come to this, we need to look in a more general way at the nature of what is created during the course of a development project.

What is created during development?

There are many answers to this question, depending on one's point of view. In Section 5.5, the point of view is one of efficiency, and there we shall use the degree of definition of the project, the reduction in risk and the completeness of the project as measures of the improvement in the result during the course of the project. In this section we shall look at things from a specification and synthesis point of view.

In an idealised project, our considerations have their starting point in the need – i.e. on the basis of the company's objectives and strategies we search after new business opportunities in the form of recognised needs. This can be a question of a more or less systematic procedure with different areas of investigation Fig. 4.7a). An area of investigation may seem particularly promising and be selected for further examination. The area may, for example, be defined by marketing or technological factors (Fig. 4.7b). Most commonly, the marketing factors will be considerably more concrete than the technological ones. Let us explain the sketches in the figure more fully:

a) The areas of investigation may be related to particular technologies or marketing channels, they may originate in accessories for existing products, or they may cover alternative uses for existing products.

b) If the search gives results, then these results may include well-defined information about needs, ideas about the type of product and a few facts about the production. The aim of the following phase, the product principle phase, is to define the way in which the product will be used, and its general principles, so that a detailed specification for the design can be worked out.

c) In the product principle phase, the production of critical components according to some particular principle may be considered in detail, but production-related factors will not normally be considered until later.

d) In the product maturation phase, the market volume must be determined as well as possible, and the product must be defined completely in the form of a preliminary design which permits the production of a working model. In this phase, production principles are also defined.

4.7

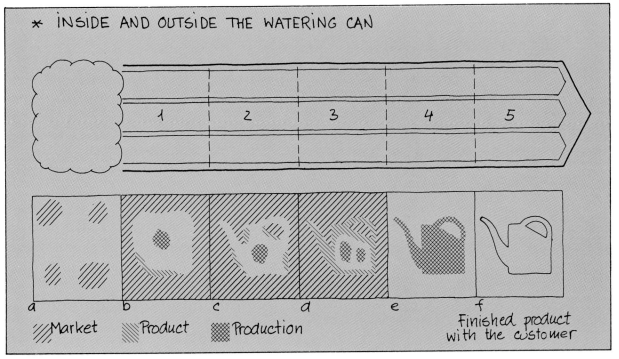

* INSIDE AND OUTSIDE THE WATERING CAN

a b c d e f

Market   Product   Production

Finished product with the customer

e) In the production preparation phase, full details of both the product and its production are fixed, so that only a few unnoticed details or features of the running production remain unknown. Preparations are made for sales – i.e. the sales apparatus is defined, leading more or less to completeness in the marketing matrix.

f) In the realisation phase, specifications are no longer of interest, because we move from the speculative and risky world of preliminary development to the world of reality: with the finished product in the hands of the customer.

In this sequence of events we notice the following:

- In certain phases, the task of development is principally one of specification – i.e. results are created (based on decisions) in the form of objectives and strategies, and later on in the form of demands, criteria and requirements. This is the case in the specification of the business, the product, the market and the production.

- In certain phases, the development task is mostly synthetic – i.e. results are created which correspond to or complement the specification. This is the case for the market, the product and the production system.

- There is often an overlap between the activities of specification and synthesis. We are never in a 'pure' situation, but regularly specify things in terms of things which have already been decided on and realised – in the case of products, some typical examples are the principle of the product, its structure, design and shape.

- The difference between very innovative and less innovative projects does not lie in the sequence of activities, but in the extent to which things are predetermined.

This view of product development, which we call 'the watering can seen from inside and outside', can be used to define various types of project, categorised by the number of decisions about their realisation which they involve.

The point is that all projects start with a rather definite degree of concreteness, i.e. definition of the 'internal watering can'. It is important to recognise this start state, as it defines the extent and type of the following development process, and the extent to which it is possible to create something new (Fig. 4.8).

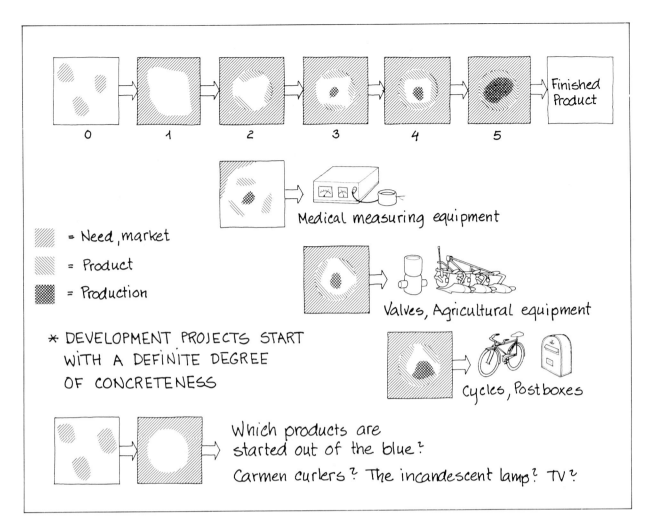

= Need, market

= Product

= Production

* DEVELOPMENT PROJECTS START
WITH A DEFINITE DEGREE
OF CONCRETENESS

Medical measuring equipment

Valves, Agricultural equipment

Cycles, Postboxes

Which products are
started out of the blue?

Carmen curlers? The incandescent lamp? TV?

4.8

Design degrees of freedom

As mentioned on page 24, we know how the activity of design proceeds. It can be described by the so-called design characteristics shown in Fig. 4.9, and the sequence of events consists of the successive definition of these characteristics. Complexity increases on the way, and many alternative solutions may be associated with each characteristic. Thus, if we picture the course of events as a pyramid, then the pyramid must be multi-dimensional. A concrete solution arises when we define a particular process (for example, a measurement) that the product must carry out, describe which functions must be present (e.g. receiving a signal, amplifying a signal), and find principles which can be used for the realisation of each function (e.g. a piezoelectric effect, a voltmeter). Such principles can be fitted into spatial structures in many different ways. When a particular solution has been chosen, we start to define the mechanical parts and their relationships to one another; during this process, the form, composition, dimensions and surface quality of each mechanical part is determined. At this stage, the design task is finished.

The starting point for a project may be an open problem, as in the example of Fig. 4.9, or it may lie some way down in the pyramid, corresponding to the principles of the product already being fixed, or the design of certain components already being fixed. In the

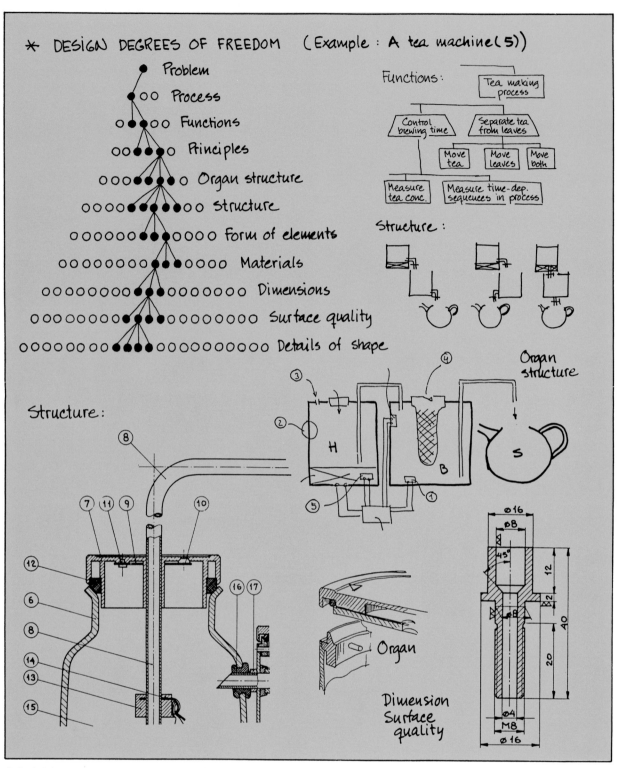

DESIGN DEGREES OF FREEDOM (Example: A tea machine (5))

- Problem
- Process
- Functions
- Principles
- Organ structure
- Structure
- Form of elements
- Materials
- Dimensions
- Surface quality
- Details of shape

Functions:

Tea making process

Control brewing time — Separate tea from leaves

Move tea — Move leaves — Move both

Measure tea conc. — Measure time-dep. sequences in process

Structure:

Structure:

Organ structure

Organ

Dimension Surface quality

4.9 (after(5))

literature, there are many different concepts used for characterising project types, for example those shown in Fig. 4.10:

- Updating/replacing, corresponding to the case where existing products continue on existing markets, i.e. with the same applications.
- Adaptation, where existing products are modified to suit new areas of application, i.e. new market segments can be taken up.
- Supplementing, which means the development of new products for existing areas of application.

4.10

* PPOJECT TYPES

Products →

	Existing	New
Existing applications	Updating Replacement	Supplementing
New areas of applications	Adaptation	Diversification

Markets ↓

☐ Are called "new products"

- Diversification, which is the highest degree of innovation, in which new products are developed for new applications.

These concepts are not particularly self-explanatory, and may lead to confusion. The concepts used by Roth [12] follow the ideas shown in Fig. 4.9, and his concept of 'new design' corresponds closely to the concept of 'new products' used in the American references cited in this book:

- 'New design', i.e. one product development type in which the formulation is open, and where the functions are not known in advance.
- 'Function design', where the functions are known, but where one or more principles in the product are unknown.
- 'Shape design', where the shape is more or less given, but where details of the structure and shape are free to be fixed according to the principles of design for ease of production and assembly. (For a commentary on this, see Chapter 8.)

A useful concept in this connection is 'design degrees of freedom', which can be thought of as 'how far up the pyramid do we go?' (Fig. 4.9) or 'how well do we exploit the alternatives included in the layers of the pyramid?'

> The starting point for the project is defined by the design degrees of freedom

In other words, it is necessary during definition of the innovation or ambition level for a project to define what features of the project are predetermined and which are free.

Definition of this starting point often causes problems, because the product is complex and certain details in the model on which the

4.11

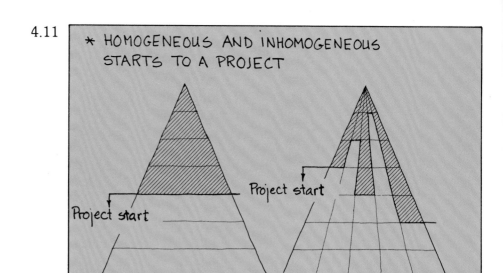

* HOMOGENEOUS AND INHOMOGENEOUS STARTS TO A PROJECT

Project start

Project start

Sub-systems

product is based are to be preserved, whereas others – for example a whole sub-system – are to be altered. Figure 4.11 shows how the pyramid can be thought of as divided up into segments, corresponding to subsystems, each of which has its own starting point.

Figure 4.12 shows that the definition of the necessary principles, structure and design are out of step in the different subsystems. When the main principles have been fixed, one can begin to consider the principles to be used in the auxiliary subsystems. The result of this has an influence on the way in which the structure is chosen, in the way that the structure of the subsystems has an influence on the form of the main systems, and the form of the subsystems has an influence on the form of the main systems. This pattern continues until we reach a level of detail where the mechanical parts can be put together. Here the game stops.

4.12

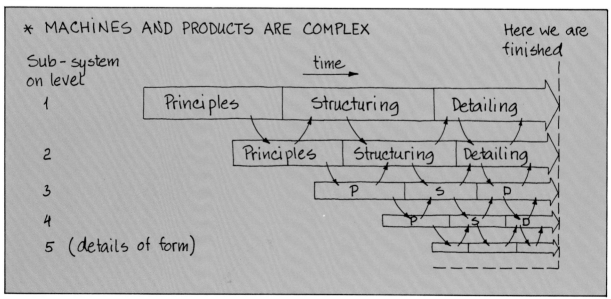

* MACHINES AND PRODUCTS ARE COMPLEX

Here we are finished

Sub-system on level

time

1	Principles	Structuring	Detailing
2	Principles	Structuring	Detailing
3	P	S	D
4	P	S	D

5 (details of form)

'Innovative modification'

A very large proportion, perhaps 80% of all product development, is design modification with the aim of updating the product – i.e. making small alterations, perhaps with small evolutionary improvements. The title of this section may seem paradoxical, but it is very important to recognise that:

Projects should be innovative all the way through

By this we mean that the design degrees of freedom should be exploited at all levels in the pyramid (Fig. 4.9) so that innovative solutions are created for subsystems, details, processes and methods of assembly.

The life of a product on the market can be divided up into characteristic phases (Fig. 4.13). Modification is typically something which turns up at the end of the growth phase and during the whole of the maturity phase. If we look at the idealised complete sequence of the search for business ideas and the subsequent development of a product, as shown in Fig. 4.5, this figure only need to be transformed very slightly, since the basic course of events is the same for all types of project. There must still be phases in which the need is recognised and screened, where one discovers which changes in the need have lead to the requirement for a change in the product, but it will not normally be necessary to prepare for development of the product – the projects can be started at once. The content of the actual task of development will of course be condensed somewhat, as the general principles of the product, its structure and to a certain extent its design, are known (cf. Fig. 4.11).

4.13

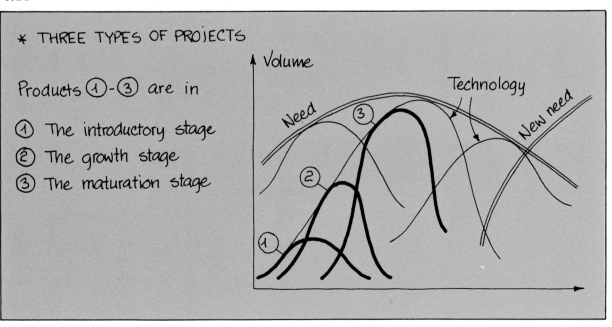

5

Getting results out of product development

SUMMARY

The management plays an important role in planning and following up development projects. The necessity and volume of product development activities and the need for and handling of ideas are important planning questions, and the problem of setting the right things in motion and setting the right people to carry them out is a central one.

When projects are underway, one needs to concern oneself with important parameters which are of significance for the creation of results; namely the ability to be innovative, efficiency, dealing with questions of finance, and making the most of the business.

5.1 Extent of product development

The extent of product development must be adjusted to match the prospects of the company, that is to say an evaluation of the company's situation reaching some way out into the future (Fig. 5.1). It can be difficult to imagine Doomsday at some time in the future, when the current situation is excellent. And the opposite is also true: if the current situation is poor, then there is no surplus of optimism which might enable one to control the future.

5.1

In product development, the necessary meets the possible

'The necessary' is the creation of business results, i.e. the attainment of short- and long-term economic targets. 'The necessary' may be the unpleasant result of an analysis of the contribution of new products to profits or of the gap between profits as they are imagined and as they exist in reality (Fig. 5.2).

5.2

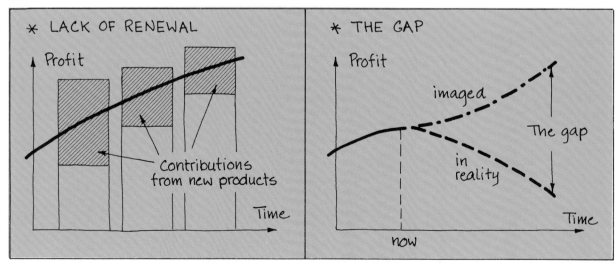

Create your future

'The possible' is the utilisation of product development to renew the company's business opportunities by development of marketing and by exploiting technology to give new business. Here a disturbing number of possibilities present themselves; by choosing among these, the company determines its future. 'The possible' is of course controlled by the market's acceptance of the products. The marketing people will claim that something 'necessary' is present in this feature of things. It is healthy if you see a compelling situation as a chance to do something quite different, to create something new.

How many projects are required in order to create some degree of renewal? If we start by looking at fundamental relationships, then we must put $n = N/L$ products into production every year, where N is the number of products embraced by the company and L is their average market life. If the company desires 10% expansion per year, then the number of projects must be $P = (n + 0.1N) \times U$, where U is the duration of the project or the development time from creation of the idea to market launch.

As an example, if the company has 10 products, each with a market life of five years, and a 10% expansion rate is desired with a development time of three years, then there must at all times be $P = (2 + 0.1 \times 10) \times 3 = 9$ projects underway. Production of a new product must be started every 4 months [18].

In the USA, the contribution of new products to company profits was 20% in the period 1976-81, and was expected to reach 32% in 1981-86, within mechanical industry [17]. These expectations of growth must be borne by technological development, changes in consumer needs, shorter product life cycles, and increased access to foreign markets. Working in a negative direction are factors such as increased capital costs, government interference and wage costs.

How many projects may be allowed to fail? How do we dimension product development so that the stream of business ideas leads to a suitable stream of projects? In Section 4.2, we saw how there ought to be a large number of ideas to choose between and to coordinate

5.3 (after(17))

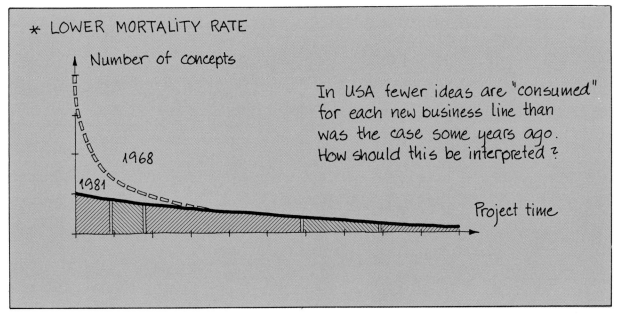

in the early phases, if good business is to come out of it. An investigation in the USA has shown that this 'mortality curve' has changed in appearance between 1968 and 1981, as shown in Fig. 5.3. Due to increased attention to market conditions and the potential borne by new applications of existing technology, the 'necessary number of concepts' has been considerably reduced.

An attempt to illustrate the dynamics of the mortality mechanism is shown in Fig. 5.4. Here, the sequence of events is imagined as three tanks containing business ideas, project proposals and product proposals, out of which there should preferably come some useful business. The activities of refinement and selection reduce the number, but the tanks are also leaky, because ideas become out-of-date, are abandoned or are chosen by one's competitors. The faster and more efficiently things run from tank to tank, the more useful business will come out of it.

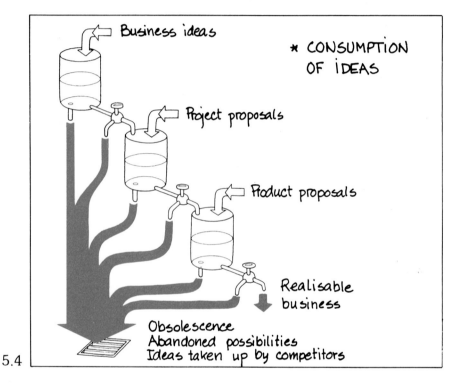

5.4

5.5 (after(17))

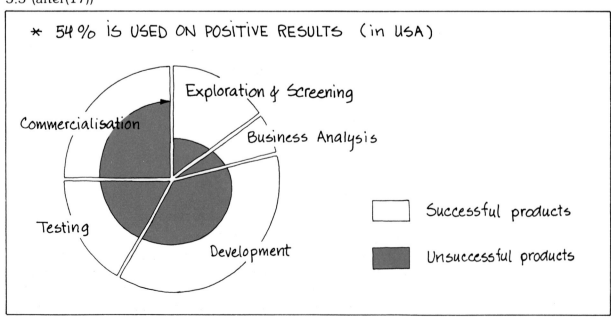

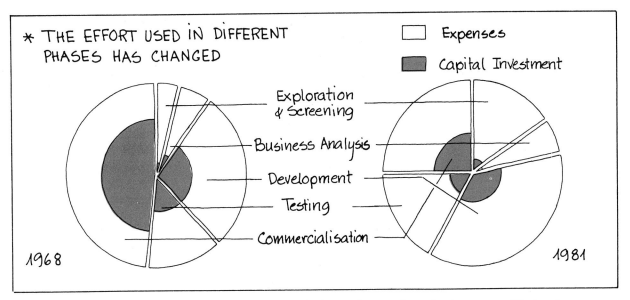

5.6 (after (17))

What is expended on things that don't succeed? Again we can quote an American result, where the average of 700 companies embracing 13,000 new products [17] gives the picture shown as Fig. 5.5. In Section 5.2, we shall examine more closely the factors which commonly determine whether the result will be a success or failure.

Finally, let us look at the extent of the individual parts of the collected product development. The way in which costs are divided among a project's various phases, as presented in Booz, Allen & Hamilton's investigations of American companies [17] is shown in Fig. 5.6, where projects are for the sake of clarity made equally long (perhaps they are carried out more efficiently today?) and the phases are likewise made equally long. (The phases are the same as in Fig.5.3.) The curves show increased costs in the early phases. In successful firms this reflects their use of specific strategies for new products, their use of more analyses, and their concentration on the creation of ideas and concepts. ...And the Japanese use even more on the early phases.

"A new – and we think important – managerial concept emerged from the survey: the effect that a consistent commitment of resources to new product development has on gaining and maintaining a competitive advantage" [17]

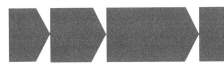 There is no cheap way to successful innovation

One should not be fooled into believing that success can be achieved just by chance through small investments in basic ideas. Everything indicates that the day of the lucky small inventor is over. Product development demands high-risk capital investment to an extent which completely overshadows the costs of creating the result, and the road to success is paved with numerous expended ideas and abortive attempts.

Product planning never holds, nevertheless planning is very important

Product development is made vastly more efficient if there is planning and coherence in the creation of business – i.e. if considerations and results can be carried over from one project to another, and if the staff get training in carrying out projects by being allowed to try it several times and to observe their strengths

73

and weaknesses. We shall look more closely at the project leader's role in this connection in Chapter 6.

5.2 Launching a project

To set a project going is like launching a ship and getting it into motion along the right course with the right crew. But the special aspect of product development is that the ship almost doesn't exist when it is launched. We hardly know which way is up and which down, what is fore and what aft – perhaps the vessel is barrel-shaped – and we shan't find out where it is going until it starts to move.

If we now want to look at the problem of setting the right things going, we must therefore look both at the start of the project and the way in which it is followed up before we can be certain that we have set the right thing in motion. Ideally, we only need to have sufficient insight into the market, the product and its production to be able to set the development of some good business going, but here too we are on thin ice because this insight must necessarily be lacking at the start of an innovative project. To continue with the boat analogy, we start our voyage without knowing the course exactly and without having a detailed chart. And the destination is perhaps just an illusion, an alluring South Sea island.

In what follows we shall in a very general manner look at the problem of starting innovative product development; the details are considered in several other sections of the book. The image of 'starting a voyage' will be used as a continuing analogy.

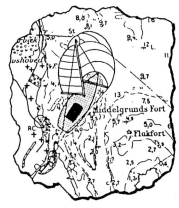

– chart the waters

Creating the chart has as its objective the identification of rocks, shallows, destinations, attractive waters – to use marketing language, we need to build up knowledge about the market and the product, the division of the market into segments, trends, niches and so on, so that the marketing strategy can be set. We make the chart for those areas which we find interesting – creating a total view is an impossible task. And as for those charts which we can buy in the shops, well, our competitors use them as well!

Product development is a progressive process in which the target moves, and where the course has to be adjusted as we go along. So it is necessary to build the chart up and fill in its details as we go along – i.e. we must obtain more and more precise knowledge as we carry out the project and the business opportunity becomes more and more concrete.

Evaluation of the ship's sea-worthiness is a condition for being able to organise the voyage: is it a coffin ship, a luxury liner or a yacht? The capacity and technological ability of the company must be appropriate to the task in hand, and there are often some logical conditions or keys to success –access to specific knowledge, experience with respect to development costs, the knowledge base of the project group, and so on – but at a more general level the management must be geared up to product development. In other words, a development philosophy, strategic thinking and a style of leadership which can get product development to flower must

– is it seaworthy?

prevail. Investigation show [17] that the following factors contribute to a new product's success (Fig. 5.7). Some of these factors are related to 'sea-worthiness', others are related to the course chosen, and others are related to the 'match' between the development task and the organisation. But of course the figure has been created in the clear light of hindsight, and can primarily be used to get a general view of what is important in a project.

The results given in Fig. 5.7 must be viewed with certain reservations, since conditions vary so much from one branch to another, and from one type of product to another – for industrial products, for example, technological dominance is the most important factor. One might enquire what it is that distinguishes the successful companies form the less successful ones. At first glance they appear to be identical: they introduce the same mixture of products, use the same procedures and have the same distribution of costs between the different phases! The difference is [17]:

- Management philosophy: The successful companies are based on growth, created by means of products they themselves have developed. They have used formalised procedures for a comparatively long period. Part of their strategy is that a particular part of the company must grow on the basis of the new products. They work through business ideas thoroughly, and consider 10 times fewer alternatives per successful product than less successful companies do.

- Organisation of development: As a rule, the successful companies have development of new products placed in a development or 'engineering' department and give their research, development and marketing functions some

5.7 (after(17))

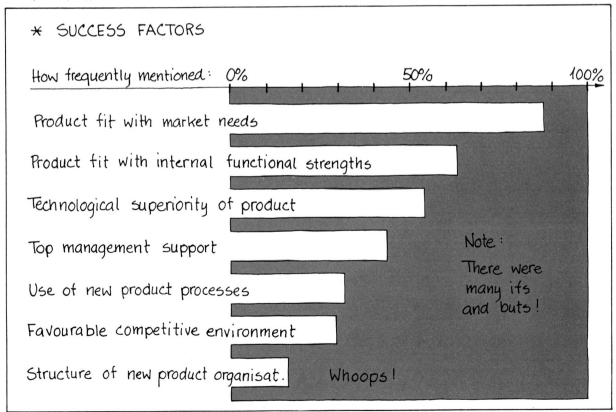

75

influence on this development. They also keep their head of development in his job for a longer period.

- The effect of experience: The successful companies accumulate experience and knowledge and continually make themselves more efficient (cf. Sections 5.3 and 5.5).

- Management style: The successful companies do not just choose a management style which suits the development of new products, they also tailor the style to suit changing new product opportunities. The following types of management appear in the most successful concerns:

  –The entrepreneurial type, in which an autonomous group is created to look after new products, and reports to the top management. This group consists of a multidisciplinary team, lead by an entrepreneurial person who understands how to integrate the various functional skills. The top management is closely involved, while less attention is paid to commercial planning and the use of financial criteria. This less restrictive attitude creates a positive environment for the taking of risks, and for project leaders who appreciate a system which gives suitable rewards to success. It is therefore possible to hold on to the project leaders, so considerable continuity and the accumulation of experience relevant to the development of new products can be attained.

  –The collegial type, which is characterised by the management being heavily involved in the decision-making process and in the taking of risks, by the management's commitment to and support for the development of new products, by formal procedures and by clear commitments across departmental boundaries in order to create what is necessary for success and the rapid taking of decisions.

  –The structured type, characterised by a hierarchic management structure with many levels and strong direction from the top. This type of company puts a lot of emphasis on the leaders of different functions, is heavily oriented towards business planning, often uses an inflexible procedure for the development of new products, and relies heavily on formal financial criteria when making decisions concerned with possible new products. This type is suited to dealing with existing areas of business, because it rewards those product managers who gain success. Nevertheless, this type appears to limit efforts to create new products to variations on existing products, and it gives less continuity than the other types, because the leaders get promoted out of their own areas.

These four conditions – management philosophy, organisation, the effect of experience and the style of management – are, as previously mentioned, important indicators for the 'seaworthy' success companies. We have quoted at length here, because we feel that these conditions can be carried over to companies in other countries, where, it must be noted, product planning, management involvement and professionalism in the execution of development projects could be given higher priority.

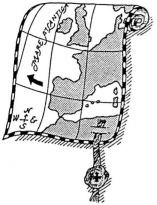

– sailing instructions

– starting conditions

If we return to our images of launching a ship and sailing a voyage, which are a parallel to the management's creation of a development project, we find several similarities:

Sailing instructions. Our voyage or product development is not a once-only experience. There are preceding and following and many simultaneous activities to take into account, and we must also allow for the ballast, which the company provides: existing marketing, product know-how, and production know-how and machinery. Taking such things into account must therefore be built into the projects, formulated as strategies for the areas of marketing, the product and its production.

Starting conditions. The voyage itself and its destination – i.e. the project's subject or the basic business idea – are naturally also important for the result: are we going out in a winter storm in a kayak, or going on a picnic in pleasant weather? The basic conditions for the start of a project are of course the existence of the external prerequisites, i.e. capacity, resources and technical ability.

There are many methods and techniques for evaluation of early business ideas [19-21]. This problem area resembles the evaluation problem at the project's key points, and will therefore be dealt with later in Sections 6.6 and 6.7.

Follow-up is, as mentioned at the beginning of this section, just as important as start control for the definition of the project's course.

The management should observe the following:

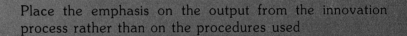

Place the emphasis on the output from the innovation process rather than on the procedures used

The management must demand guiding signals from the projects that are underway, so it becomes possible to evaluate whether they are on the right course, and whether the prospect of profitable business is still in sight. We shall return to these questions of the management's project control at the end of this chapter.

It is crucial for innovation in a company that it has the attention of the management, not just during the launching ceremony, the result phase of especially important projects, but in all projects in all phases. The degree of involvement demonstrated by the management is reflected in the staffing of the project management team, and in the signals which are regarded as most significant. In one Danish company making apparatus, the management checked the servicing arrangements for the apparatus, because this was their Achilles heel, but otherwise no attention was paid to details. In a Japanese company, which only held a product planning meeting once a year, and there took the temperature of all the projects, nobody was in the slightest doubt: one should check the market situation and the course of the project in relation to the deadlines, i.e. external conditions and internal results.

5.3 Personnel

The next most common success factor in Fig. 5.7 is the way in which the product matches the areas in which the company functions strongly, i.e. the areas in which it possesses know-how. So in this section we shall consider how we create and deal with the project group, so that its strong side is developed.

Putting the group together in the best possible way

Projects are multidisciplinary and interdisciplinary, they require insight into all areas of the company, and an overall grasp of the economic interplay between these areas. Thus it is normally necessary to let product development take place in a project group, as emphasised several times above.

Working in groups has advantages and disadvantages. The basic mechanism is that employees with different expertise in the areas of marketing, development and production create a pattern of working where their individual abilities supplement one another (Fig. 5.8). The effect of good and bad collaboration, compared with the efforts of a single person, are shown in Fig. 5.9. Communication is both a necessity and a source of costs when several people work together. Figure 5.10 shows how an employee's productivity is dependent on the proportion of his/her time spent on the project. Below a certain limit the situation is meaningless; the person's productivity approaches zero, as all the time is used for communication and the accumulated effect on the project becomes negative.

The project group normally contains members with two different tasks, namely as representative for a user/buyer/client/subcontractor or as specialist/technical expert in the project itself. (Page 80 shows the ideal qualities demanded by these roles in project work.) The project should not be put together from people with identical personalities, but should rather contain a suitable

5.8

* THE TEAM MEMBERS' KNOWLEDGE MUST BE ADDED UP

Knowledge of market

Knowledge of production

Knowledge of development/design

Knowledge of market

Knowledge of development/design

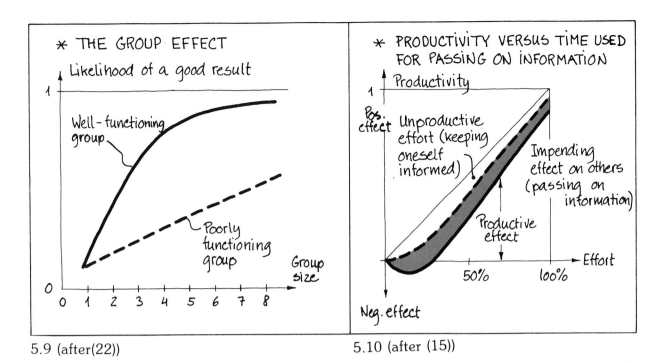

5.9 (after(22))

5.10 (after (15))

spectrum of innovators, leaders, systematists and so on. This question is dealt with in several references [15,22,23], which also discuss an advantageous mix of qualifications and personality types (Fig. 5.11):

● The creative product developer is important, but his role is over-estimated. His presence does not ensure efficient product development; you also need the types to be discussed below, who are often overlooked. Staff for product development are often recruited as if they were all expected to be creative product developers, but really only a few are able to satisfy this requirement.

● The entrepreneur, who possesses an extrovert, aggressive form of creativity who sells the group's ideas, and works from more emotional motives, while the creative product developer is more rational. The entrepreneur has to be stimulated and lead in quite a different manner from the creative product developer.

● The project leader or standard bearer, who is good at organising things, pays attention to the needs of those whom he is supposed to coordinate, and is an efficient planner. In Danish industry, there is often doubt about what technical qualifications he should possess and how he should exploit them. We shall return to this topic in Chapter 6.

● The sponsor, who is often an older, experienced product developer, who is able to act as a link to the basic organisation and can argue the case for the project before the management. His role, though it goes unnoticed, is often decisive for the success of the project. In many efficient companies, the chief engineer or head of the laboratory often naturally takes on this role.

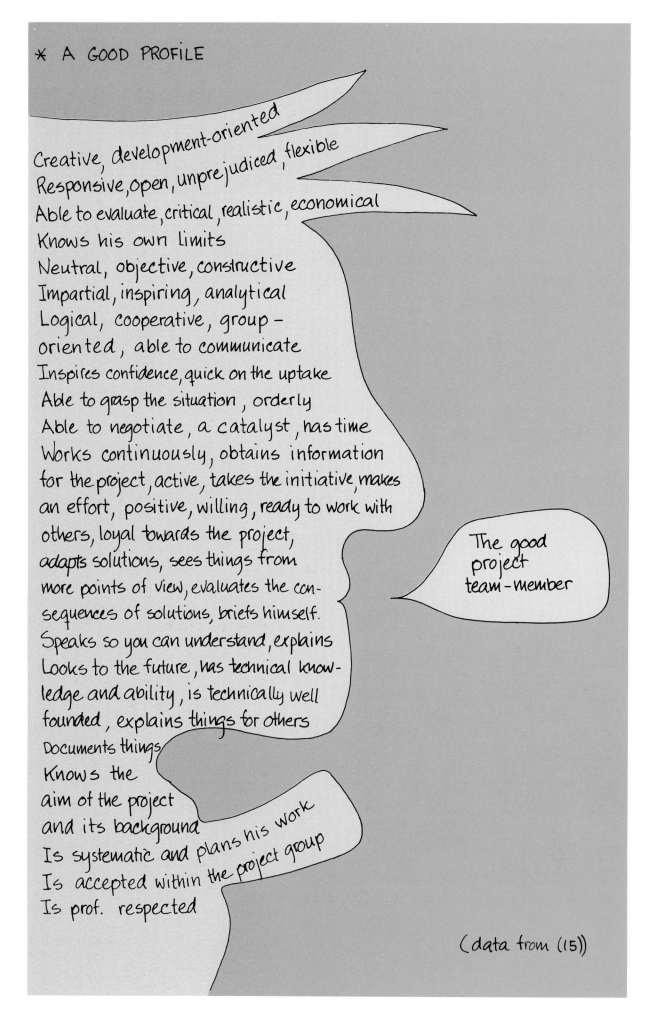

* A GOOD PROFILE

Creative, development-oriented
Responsive, open, unprejudiced, flexible
Able to evaluate, critical, realistic, economical
Knows his own limits
Neutral, objective, constructive
Impartial, inspiring, analytical
Logical, cooperative, group-
oriented, able to communicate
Inspires confidence, quick on the uptake
Able to grasp the situation, orderly
Able to negotiate, a catalyst, has time
Works continuously, obtains information
for the project, active, takes the initiative, makes
an effort, positive, willing, ready to work with
others, loyal towards the project,
adapts solutions, sees things from
more points of view, evaluates the con-
sequences of solutions, briefs himself.
Speaks so you can understand, explains
Looks to the future, has technical know-
ledge and ability, is technically well
founded, explains things for others
Documents things
Knows the
aim of the project
and its background
Is systematic and plans his work
Is accepted within the project group
Is prof. respected

The good
project
team-member

(data from (15))

80

* THE SIX

The creative product developer

The entrepreneur

The project leader

The sponsor

The information handler

The environment creator

5.11

- The information handler, who is good at finding and communicating information, and who in contrast to the previously mentioned participants reads journals and finds market information or technical knowledge. It is particularly crucial for the group to obtain knowledge of the market and insight into the competitors' situation, products and production.

- The environment creator, who is the model with respect to the creation of project culture, and who is responsible for the setting of high standards of work.

This pattern should not necessarily be seen as made up from six individuals, but rather as a set of behavioural characteristics which should be present in the project group and which can be distributed anywhere among the various areas of activity. Over this pattern we must lay the requirements for functional professionalism, which are an essential part of the basic thinking of integrated product development.

Mikkelsen and Riis [15] suggest using an evaluation profile for a project group, as seen in Fig. 5.12. The result of marking this chart should be that you can put a cross by 'good' or 'very good' for at least one person for every property, that the creative average is at least 'satisfactory', and that the group's intelligence corresponds on average to its creativity.

To what extent should the three main areas of activity and the many areas of interest be incorporated into the various phases of the project? Basically, the three main areas should exhibit more or less equally good results, but this does not need to mean that they

81

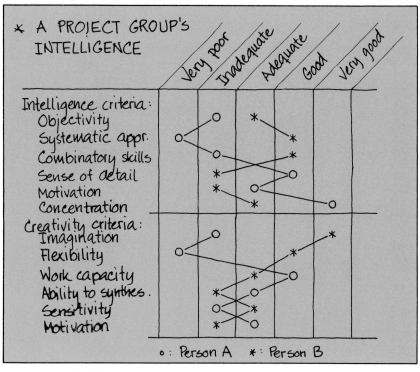

5.12 (after(15))

require equal amounts of effort. 'All projects are different' and they should be dimensioned according to this principle, but in general it is true that:

> Marketing and production should buy themselves in in the early phases of the project

Quantitatively this means that resources must be reserved for this purpose. It is not pressure from the development activity which is supposed to control this; instead the motivation should come from a desire to obtain influence and to improve one's own conditions. Qualitatively this means that people, strategies and targets which can be utilised in these early phases must be developed within these activity areas.

Finally, let us remember that project work is two-dimensional:

> The project group must be professional both technically and with respect to carrying out projects

This means that a group composed of technical specialists – however competent they are – will not give good project results, if they do not possess the quality of project professionalism. (This book deals with project professionalism!)

Creating an experienced development staff

It is generally accepted that if one frequently repeats an operation or an action, then the effort and the costs associated with doing so are reduced. On a double-logarithmic graph, the relationship

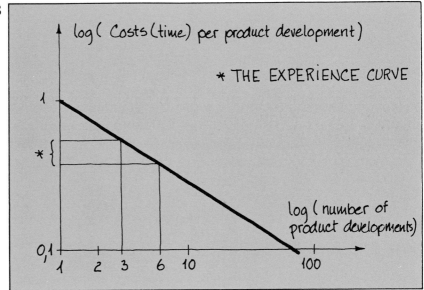

between number of operations and costs is linear, and is known as an experience curve. If we compare two points on the curve, which for example correspond to experience today and to when we have carried out four times as many projects, or which correspond to our knowing that company B has carried out four times as many projects as A, then there is a considerable difference in costs in the two situations (see Fig. 5.13).

Building up experience is the responsibility of the management

It is the management's job to arrange and organise product development so the effect of experience is as great as possible. To make the situation which we have experienced in a number of companies more easily recognisable, let us turn things upside down. Here are some suggestions for ensuring that the effect of experience is absent:

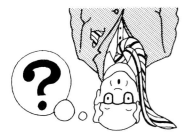

- Use inexperienced employees for product development – they are creative and have most time.
- Don't evaluate how the project went when it is finished – we all know why it didn't go perfectly.
- Keep changing the staff allocated to the project, so that the project can be seen with fresh eyes.
- Make sure that a large number of people only use a tiny fraction of their time on the project.
- Make sure to select the best project leaders and promote them to department leaders – they are sure to be good at that as well.
- Make sure that the project leaders are also involved in many other activities within the company, for example managerial tasks.
- Avoid writing down tasks, alternatives, choices, criteria, reasons or anything else which might have significance for following projects. Nobody would bother to look them up in any case.
- Don't bother to establish any specific method or procedure – make a virtue out of changing methods.

● A WORD ON BOUNDARIES

A product has many boundaries: in its origin, in its design, and in its life-cycle. In the very nature of things, product development must involve transitions from discovering a need to realisation of the product, from functional specification to materialisation, from quality requirements to designed-in quality, from production specifications to the produced product, from the 'ex-works' product to the product as sold, and, the most important threshold to cross, from the manufacturer's product to the customer's product.

Each of these transitions or boundaries represents a significant change with respect to 'what has to be done', technical knowledge, standards of values, methods, etc., and therefore offers a potential risk of the project turning out differently.

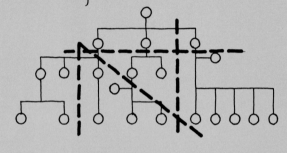

Watch out for boundaries in the organisation!

Watch out for boundaries in the course of the project!

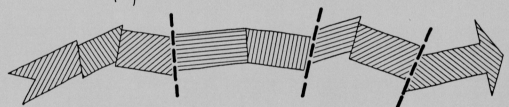

The product in itself also contains boundaries in a number of respects: between electronic and mechanical parts, between designed areas and 'the rest', between in-house production and subcontracted parts, between sheet metal work and milled parts, between 'the way we do it in Department A' and 'the way we do it in Department B', between critical parts and non-critical parts, etc.

Each of these boundaries can easily become a no-man's land or even a battle zone over disagreements.

When the project is being set up, you should take care that there are no organisational, personal or responsibility-related boundaries which are identical with boundaries or transitions which naturally arise in the product or its life-cycle.

Watch out for boundaries in the product!

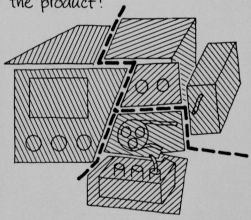

Avoid having the same boundaries within the organisation and within the product / project!

(Idea : Hans Mikkelsen (15))

84

- Reduce product development when things are going well – it's no problem at all to get started again when it becomes necessary.

What is the effect of the experience curve, and what is its slope? The American study discussed above [17] shows that the costs associated with the introduction of a new product fall by 30% every time the number of introductions is doubled. Your company can also achieve this, provided you don't give nods of recognition to the points above!

Organising on the project's terms

We have previously examined a model of the idealised project organisation (Fig. 4.1, page 52). Let us now look more closely at what the management has to consider when setting up an organisation.

5.14

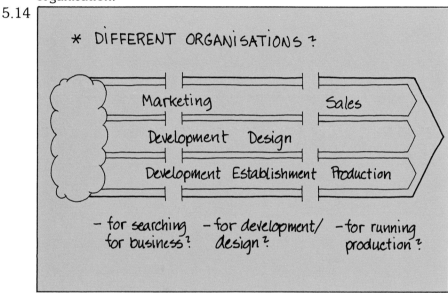

In principle we need continuously to change the organisation to match the nature of the task in hand. This is why loosely structured companies and groups find it easiest to introduce innovation. Figure 5.14 shows that we can divide development activities up in different ways and identify the main task, and thus the ideal organisation:

- Marketing involves two radically different activities. First, a strategically long-term activity in the business search phase, characterised by being outward-looking, innovative and dynamic, by an experimental attitude, by a desire to map out the ground and by searching for what to do. Secondly, in the establishment phase there is an activity which is more characterised by stability, organisation and routine, but which is still dynamic an outward-looking.
- Sales activity is dynamic and outward-looking, but at the same time stable, systematised and organised.
- Development (including, where necessary, research) involves two radically different activities. First, a strategically long-term activity characterised by searching, experimentation, innovation and specification. Secondly, a development and design activity which is more organised, systematised and short-term, and which is both subject to financial controls and capable of creating financial rewards.

● PRODUCT DEVELOPMENT IS LIKE DUCK HUNTING...

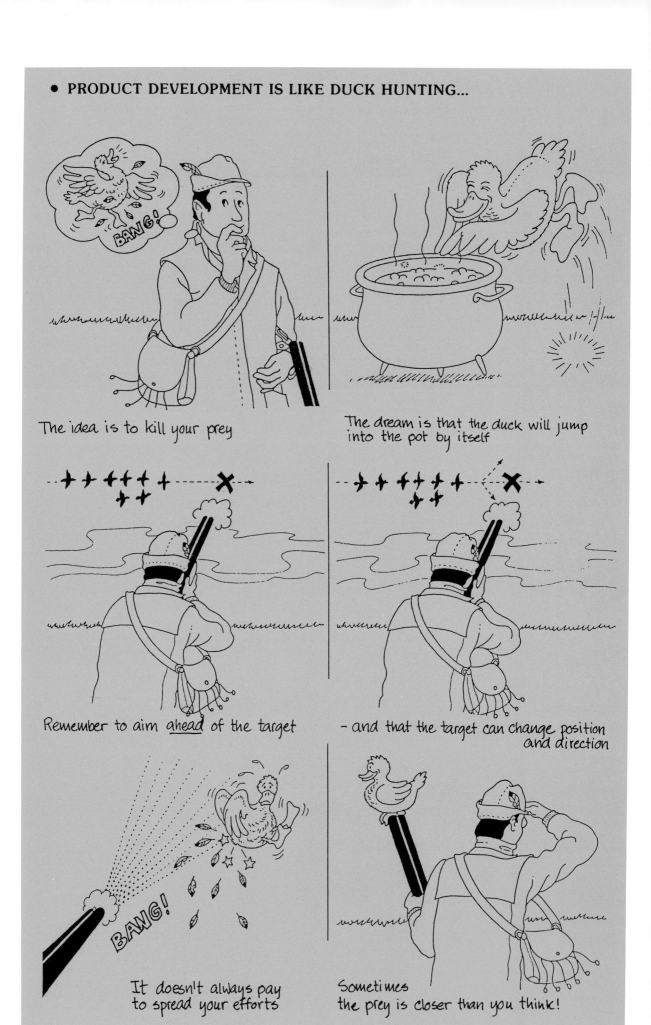

The idea is to kill your prey

The dream is that the duck will jump into the pot by itself

Remember to aim *ahead* of the target

– and that the target can change position and direction

It doesn't always pay to spread your efforts

Sometimes the prey is closer than you think!

- Design cannot be separated from development. In the production phase, it is an activity characterised by conservation, registration, routine, and the maintenance of quality, and it involves determining and maintaining the product as specified, and the execution of an important dynamic 'fire brigade' task in connection with the establishment and running of production.
- Production involves three radically different activities: a long-term strategic development activity characterised by experimentation and innovation; a more short-term design and establishment activity (the production and running-in of equipment) which is systematised and controlled by financial restraints; and a very short-term production activity characterised by a rigid organisation, with much attention to quality control and efficiency.

The people which we need most in these many different sorts of activity are of very different types, depending on the timescale of the task, the level of innovation and the degree of repetition. Thus it is important that the management knows each employee's profile, so his level of ambition and his experience can be matched to the task involved. A development project involves all the tasks listed above, and cannot tolerate changes in responsibility during the course of the project. So it is important that some members of the project team continue with the project from start to finish; on the other hand, participants with special abilities required in certain phases may of course come and go within the project group.

> **The organisation of the project group must be created on the project's terms**

... not on the terms of the basic organisation. It should never be a matter of routine to establish a project group and start a project. Superficially, the tasks involved often appear to be the same, but the basic assumptions will normally be quite different (Section 6.6). The creation of each new project should therefore be carefully considered and the project should be explicitly designed (see Section 6.4), and possibly redesigned as time goes by:

> **The project organisation should be adjusted during the running of the project**

Efficiency

The management has a noteworthy responsibility for the internal efficiency of the project group. Some typical obstacles to efficiency are:
- Imprecise and non-motivating information about the purpose and aims of the project.
- The project being given too low a priority by the management (participants often get out onto other tasks, the management takes no interest in the project).
- The project being too poorly manned, both qualitatively and quantitatively. Those who can be done without or who have 'little on their plates' in a department are put onto the project.

5.15

• The members of the group having other aims and ambitions than those which apply to the project.

These considerations indicate the necessity of careful project design (Section 6.6), following up, and intervention by the management during the course of the project.

How are projects manned in other countries? Booz, Allen and Hamilton [17] point out that the time for which 'senior new product managers' are employed has risen from 2.5 to 8 years during the last 15 years, as a result of the top management's desire to create greater continuity in product development. But paying people in this category an appropriate salary is a difficult and unsolved problem. In the USA, companies are not very procedure-minded, so they require 'the right man for the job' as a project leader [25].

In Japan [24], there is a universal opinion that those who are involved in product development must know all the company's activities and have considerable experience. So employees are not used in product development until they have 7-8 years' experience. They are full-time employed on the project, and are not placed in the group to look after particular functions, but are there to solve the composite task of creating business.

A Dutch source [1] stresses that young employees should take part in product development work, because they are more creative and not so bound by convention. But older, more experienced employees make up the main strength of the team.

In certain Danish companies, it is the case that 'the best one draws the lines', that is to say they have recognised and been able to keep the highest degree of design competence within the design department (Fig. 5.15). How many heads of R&D have a drawing board? Or a laboratory coat?

5.4 Innovation

The degree of innovation of a company should be regarded as a measure of its creative ability or potential, analogous to the acceleration ability of a car, the current capacity of an accumulator or the productivity of a writer.

In this section we shall look at the creation of an innovative environment within a company, and in particular at the role of the management in this.

We need innovation, but what is it?

Innovation means renewal. In an industrial company, innovation can arise in many ways, for example with respect to markets, organisation, routines and so on, but normally the concept of innovation is associated with product development in the sense of creating something new and developing a business out of it. The concept of newness is essential to the use of the definition: is it a question of newness as seen by the company or is it new as measured by the yard of the world outside? If the company gradually stops being innovative, it may be because the yardstick is shrinking! So the concept of newness must be related to those who know best: the customers and the market.

Innovation and creativity are often thought of together; sometimes they are even considered as synonyms. This is unfortunate: creativity is a property of persons and actions, innovation is the result of a product development effort. It is by no means sufficient to be creative in order to be innovative – you also need product development professionalism in the marketing area (recognition of needs, creation of customer contacts, business sense), in the product area (design insight) and in the area of production (know-how with respect to production technique, experience at running a production, purchasing experience, etc.).

Thus you should not expect to solve problems of innovation by promoting creativity. You need much more than this, but of course if nobody is creative then nothing happens at all. Creativity is normally defined as the ability to create something new by combining known things. This definition is somewhat lacking: New for whom? Should it also be useful? Based on what resources?

The concept of creativity becomes more operational if it is based on an attitude; namely, the attitude that one can create something, and that one dares to do it. Creativity is promoted by the change in attitudes which arises when a person has experienced a couple of times that it is possible to create new solutions, and has seen them realised.

New solutions are always based on combinations of things which are already known. The more things you master, and the more you have insight into, the more possibilities there are. The more ideas you produce, the more new combinations can be created.

How is the quality of an innovation to be defined? How can one compare productivity in different companies with respect to innovation? Patents and licences do not give a good picture, because their significance varies strongly from one branch to another. Studies of American conditions [17] show that new products fall into categories as shown in Fig. 5.16, according to the degree of newness within the company and within the market. Of the 30%

5.16 (after(17))

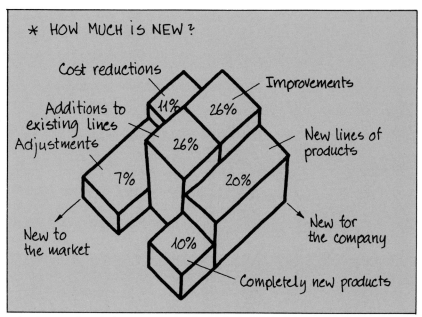

which are the most risky for the company, 60% are also the most successful. Of the companies investigated, 50% had not introduced completely new products within the last five years, and 25% had not introduced completely new product lines.

If we look at the size of the company in relation to its degree of innovation, then the American studies give the surprising picture that very large and very small companies are the most productive (Fig. 5.17). But the literature does not say anything about comparative number of innovations or efficiency.

5.17 (after(17))

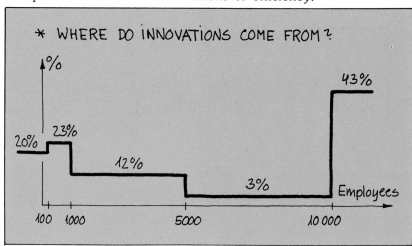

Innovation strategy

We shall not tackle the subject of strategic management here, but some general remarks are appropriate. The spectrum of strategies for creating or possessing innovative results stretches from purchase of successful small companies to the type of carefully managed and executed innovation which we find in large companies (e.g. 3M). Where one ought to try to place oneself in this spectrum depends on general factors such as the staff available, company objectives, and resources, as well as the type of market and type of technology.

Innovation can be considered as the source of energy which is required for the growth of a company. If you cannot create this nourishment, then if you are a large company you must 'consume the small ones who can, or else make do with the remains of other people's meals'. The spectrum of strategies may include investment in small high-tech companies, the creation of R&D centres within the company, the delegation of development to several small development groups related to parts of the market, the purchase of development results from special development companies or agencies, the opportunistic staking of resources on copying successful competitors etc., etc.

A common feature of all these ways of going about it is that a clear strategy is required. You have to decide what you want and in this way create your own future. To keep going is essential – you can't create results in the course of 1-2-5 years, but need patience to wait 10-15 years if you really want big results. There are no general strategies. Large successful companies in the same line of business make use of wildly different strategies.

Realising an innovation strategy

When a company decides to go in for innovation and product development, you often see some rather unhelpful remarks about exploiting creativity and making the process efficient. But where in fact do you have to make an effort?

5.18

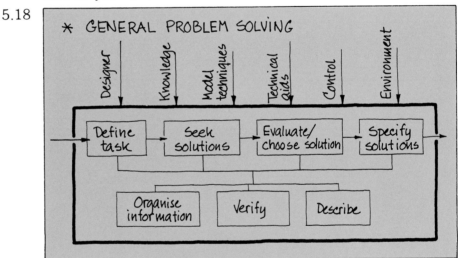

We have stressed several times that we do not believe that the form of organisation is a key to creating results, but consider the procedures used to be essential. The basic procedure in creative work is general problem solving [3], which can be modelled as shown in Fig. 5.18. The figure shows the factors which are normally of significance for solving engineering problems, that is to say:

- The designer's person and personality, with all the ways of going about things, abilities, skills, attitudes and standards which he/she possesses.
- Knowledge in the broadest sense, i.e. the know-how and accessible knowledge which one has available, or which one can discover in the course of the problem solving activity.
- Modelling technique, by which we mean the insight and the software/hardware in the form of experimental technique etc., which is available during the project.
- Technical aids, i.e. the equipment which is used in the problem solving process itself, such as drawing machines and CAD systems.

● A PICTURE OF A CULTURED PROJECT...

The idea of culturing a project doesn't need to be something dark and intangible, we can just compare it with those traditions which are prevalent in horticulture, where they have been created in order to give results.

A horticulturalist knows that trees growing wild give a poor yield, both as regards quantity and quality. So he cultivates them: by selection, by careful planting, by tying them up, pruning, watering, giving fertiliser, and so on. By creating order and by acting consistently, even with respect to small details, an optimum result is obtained.

It can be claimed that the comparison between project culture and horticulture is false, as we make an effort to extend things to their limits, and to seek the wildly growing, the creative and the risky for use in our product development. However, the horticulturalist's plants also have something of the same nature, as they are ready to burst out of their frames if he doesn't keep them in check – albeit with the aim of obtaining maximum yield.

In a development organisation this means that suitable support and room to manoeuvre must be available for creative work, but that the environment and the group must ensure that there is innovation – in other words, they must ensure that things are not just created, but also that the result is realised and marketed.

92

- Control, i.e. those activities and methods which are used to ensure control over results and resources.
- The environment in the broadest sense, i.e. the localities where the work takes place, the project group, the company, the country's laws and legal system, the patent system, engineering tradition and ethical rules.

From all this we can isolate some factors which affect the company's innovative capacity – and a large residual group of factors which do not themselves ensure innovation, but which form its basis. The most important factors affecting the individual employee's innovative ability are shown schematically in Fig. 5.19. Roughly speaking, one could also put the group in the centre, and let the figure illustrate the factors which influence its innovative ability.

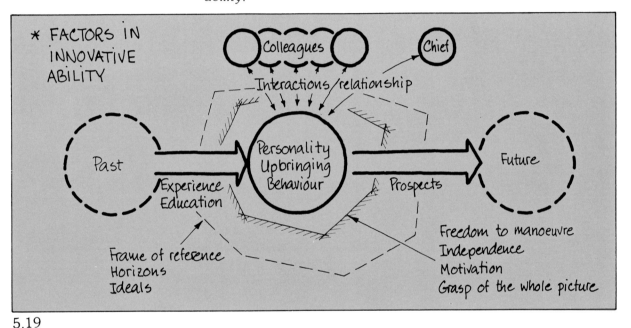

5.19

How can we influence a group so it changes 'attitude or personality' and becomes innovative? This is a question of making demands and breaking out of fixed frameworks. The objectives and level of ambition of the project must increase our efforts and show how far the group can go, and must form a challenge to everybody in the project. The project group must break the ties which inhibit innovation and must exploit the design degrees of freedom which are inherent in the problem (see Section 4.3).

Project culture

By project culture is meant the sets of values, standards and attitudes which are possessed by the participants in the project. Project culture is of great importance for important factors such as efficiency, creativity, flexibility and dynamics in the running of a project. One attempt to list the ways in which project culture, good or bad, manifests itself is as follows [26] :

- Independence and a sense of responsibility on the part of those taking part; this can lie anywhere on a scale from great energy, enthusiasm and the independent formulation and solution of problems over to stagnation, the missing of deadlines, illoyalty and misrepresentation of the facts.

- The legality or acceptability of the way in which things are managed; this is reflected in respect and acceptance at the positive end of the scale, and hidden opposition and attempts to thwart or ignore the management at the other end.

- Solidarity amongst the project participants: cooperation and loyalty as opposed to self-optimisation, egoism, isolation, covering up for oneself.

- Project ideals and standards: common, high standards for achievement and ideal working methods (care, resource-consciousness, orderliness, punctuality etc.) as opposed to slovenliness, confusion and over-specialisation.

- Coordination and cooperation: which includes efficiency, organic cooperation, mutual help and inspiration and making common plans, as opposed to self-optimisation, helplessness, negativism and ego trips.

- Openness and sensitivity to the surroundings: flexibility to changes and new demands made on the product, openness to what the client wants and to the company's overall situation, as opposed to isolation, creation of a private universe and distrust from the surroundings to the project.

- Standards of performance: orientation towards results, diligence, quality, willingness to make an extra effort, as opposed to slackness, indifference, only doing one's duty, minimum performance.

- Objective-related decisions: decisions in harmony with the overall objective, intimate cooperation, with a global optimum as target, as opposed to decisions made out of sheer need, where financial targets, resource allocations and deadlines are allowed to slide.

These manifestations of project culture show clearly that it is difficult to draw the line between culture and professionalism; it can be claimed that some of the negative points indicate lack of professionalism. But it is important to remember that culture is a property of the company (although borne, of course, by its employees) and that it can be and must be cultivated. Professionalism can be studied, learned and trained, and is much more absolute in its nature.

We need both professionalism and culture

It is clear from the list above that project culture is very dependent on the social and human relations between employees. It presupposes mutual acceptance, respect and cooperation (but not necessarily unanimity), friendly opposition, fellowship, enthusiasm, and so on.

One obtains a good idea of a company's culture if one as an outsider gets to hear the myths and off-hand remarks going round in the development department when they discuss the marketing or production departments, or vice versa, when one hears how the ordinary employees characterise the management and its way of running things, and the ordinary employees' interpretation of objectives and results.

The management's role in the creation of innovation

A management which does not involve itself in product development has no positive influence on the environment for innovation. Figure 5.20 shows the surprising result of a study of management activity in a project compared with the significance of the decisions [23].

It is the task of the management to encourage project culture and thus to make the company innovative and efficient. This requires that the employees' attitudes have to be influenced and changed. Thus,

> The management must set a good example

Giving a good example with respect to how to deal with things and how to behave is a powerful means of influencing others. This presupposes, of course, that the employees come into contact with the management in connection with crucial questions and problems. In many companies, this contact is completely absent.

5.20 (after(23))

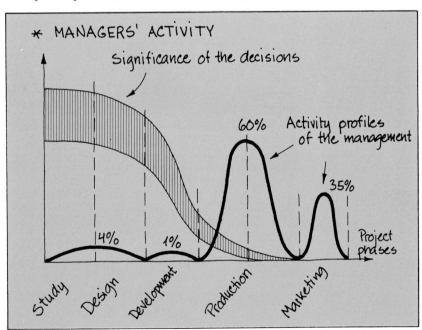

> The management must provide motivation

The delegation of power does not automatically imply motivation. More is required: enthusiasm, orientation towards objectives, esprit de corps – not in the individual department or group, for there it can do more harm than good, but across the organisation, related to the development targets which are to be attained.

The physical organisation of these forms of interplay is not unimportant. The staff must be collected together, there must be a high concentration of them where development activities take place, the localities must be such that they are clearly identified with an innovative environment and as a place where results are achieved, they must be suitable for the purpose and the distance law (see Fig. 5.21) must be obeyed.

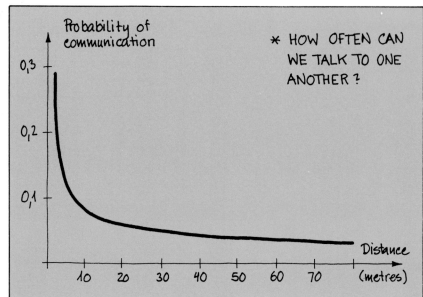

5.21

The cultivation of innovative capabilities within an organisation requires training, that is to say selection of those who have the necessary professional requirements, motivation, training and finally effort, exactly as with a football team. In-house courses and training of complete groups, initiated and monitored by the management, are much more productive than individual, 'reward' courses held by external training organisations.

There is a need to create opposition

It is impossible to use or increase one's strength if one's surroundings do not offer challenges, make demands or provide opposition.

Many companies are characterised by having a 'vacuum' round product development, and therefore only attain mediocre results. If the boss is the only one to formulate a project or a task, then it never gets tried and tested through being opposed by someone who thinks and reasons in another way. If one department has most to be said when it comes to product development, then its influence and decisions become toothless if no opposition is created which can bring about discussion and reflection. In the establishment of project groups, it is just as important to find good opponents as it is to find people who 'get along'.

5.5 Efficiency

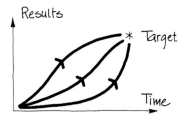

The aim of project control is to ensure that results are attained in accordance with formulated objectives. But how do we control things, so we not only get our mail on time but also the quality we want on time? The problem can also be put in this way: how do we ensure that the amount of work consumed is exploited in the best possible way?

How is efficiency in product development defined? Presumably not as the number of drawings per month or the number of new products per year. We must take the nature of product development as our starting point, and ask what it is that alters as time goes by, and which distinguishes the situation at the end from that at the start of the project. Of course, we have some finished drawings for a product which was unknown at the start, and we have established a production system which can produce the product. But is is more significant that we have clarified a long series of matters: we have interpreted the market, chosen the correct performance, chosen the correct form for the product and created a product at the right price, and we have clarified how production and assembly processes must run in order to achieve a successful production. And it is even more significant that we have proved that our choices are the right ones, i.e. that we have removed part of the risk: we have become more knowledgeable.

Efficiency can thus be defined as:

$$E = \frac{\triangle \text{clarification} + \triangle \text{risk reduction} + \triangle \text{detail} + \triangle \text{documentation}}{\triangle \text{costs}}$$

Sometimes there is an increase in all three quantities, in other phases only one of the parameters increases. There are even project activities where we 'don't get anything for our money', for example where we wait for the completion of a working model or make a clean version of the drawings.

Product development with its key points can thus be considered as a process where the three efficiency parameters gradually increase as shown in Fig. 5.22. The art of carrying out a project is to arrange for the correct rate of increase in the creation of results. If you make too slow progress, for example by undermanning a project, then the project becomes a 'meetings and folders' project, and nothing is created. If you go forward too fast, for example because 'it is nice to see concrete results on the table', then the project group is obliged to skip things and to make risky decisions, and the project's internal degree of risk may become too high (Fig. 5.23).

5.22

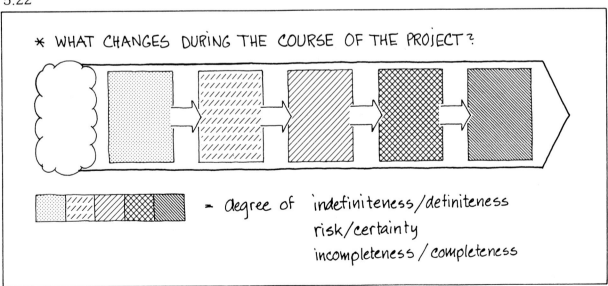

* WHAT CHANGES DURING THE COURSE OF THE PROJECT?

- degree of indefiniteness/definiteness
risk/certainty
incompleteness/completeness

How can efficiency in a project group be increased? We need to look at the three efficiency parameters and at the factors which affect them.

* WORK ON THE PROJECT IS RUINED BY
- meetings and files
- too rapid decisions on details

Basic specification

5.23

The degree of clarification in a project implies a movement from the indefinite and undefined to the defined – to something one can grasp and which one knows something about. Recognition of this is most often based on unsuccessful attempts and incorrect solutions. The following factors are significant:

- Ability to see the entire range of solutions, i.e. to grasp which solutions one chose/did not choose, and which ones the competitors chose. In this task, experience and the use of systematic methods play an important role.

- Knowledge of the parameters, i.e. knowledge of which parameters are significant for the performance of the chosen solution, and how critical or sensitive these parameters are with respect to changes in details of the product, production conditions or conditions related to the method of use or the user environment.

- The breadth of the development activity, i.e. the mass of iterations, poor shots, detours, alternative searches, attempts to follow other routes and so on, which have characterised the project. All this contributes to our understanding of the solution which in fact was chosen. Good people learn from both 'correct' and 'incorrect' activities.

- Spin-off, both from the current project and from the other projects which are in progress. Transfer of knowledge and experience to/from other projects is significant (so we don't proceed in too narrow, objective-oriented a manner) and contributes to the collective efficiency of the organisation.

Risk reduction involves the removal of uncertainty, ignorance and unknownness in project activity. As we shall see in Section 7.4, it is important to handle the risk in a project. The most important means for doing this is of course:

- Proof, that is to say realisation of the speculative solution and testing it against the right conditions, so that the hypothesis is checked. We can prove things by using working models, prototypes, shape models, pre-production runs, marketing trials and so on, and by exposing these results to a more or less 'real-life situation'. (A research laboratory is not so realistic as a field trial, which in turn is restricted in relation to use of the product by a customer.)

Detailing involves making things complete, quite finished, elucidating all the details and including them in our considerations. The following factors are important for this:

- Structuring, i.e. obtaining a general grasp of the product's components and the project's sub-activities, and ensuring that they are well-defined. Knowledge of the significance of the structure for costs, which will be incurred in the realisation phase (assembly, flexibility, production cycle time, the feasibility of related projects).

- Detailing, i.e. professionalism in making the detailed design, and in defining the production and marketing processes in detail.

- Handling interfaces in the broadest sense of the word. Embryonic problems are created in the interfaces between electronic and mechanical parts, between marketing and design, between packaging and transport, between this product and the equipment the user already possesses, and so on – because several groups of people divide the responsibility for these interfaces.

What is an efficient organisation? If we compare this idea with the concept of an efficient designer, then we are struck by things such as overall grasp of things, experience, vigour and professionalism. How does the 'experienced organisation' work? As mentioned previously, a smaller 'consumption' of ideas has been observed in the USA today compared to the situation 10 years ago (Fig. 5.24). Perhaps this is the sign of a professional organisation which creates interplay between development projects, which overall launches more new products (thus exploiting more of the early ideas) and which perhaps is technology oriented (and therefore does not to such a great extent 'start from a need situation').

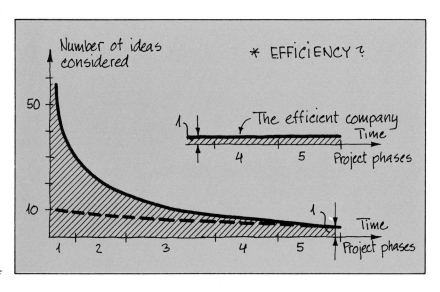

5.24

A boundary case is the efficient company which carries through its ideas, perhaps not always at very high levels of ambition, but always reaching a result. Or could this be the picture of the unprofessional company, which is unable to handle the early phases, grabs a chance idea and carries the project through at all costs, since heads would roll if it failed?

5.6 Financial control

The financial aspects of a project, together with deadlines and quality, make up an important measure of results, and financial control is a central aspect of product development. Financial control without a target is impossible, but as we shall see in what follows it is not enough to aim for the 'lowest possible costs', you also need to aim for quite definite, optimal cost structures.

Project costs

The costs involved in a development project can be shown diagrammatically in a so-called cash flow curve, i.e. an overall picture of returns minus fixed costs and annual investments. A project – and in particular the period in which the initial investment is regained – will run for many years, and we all know that income today is worth more than income in the future. If we take the total depreciation of capital to be about 30% per annum (including inflation, bank interest and internal interest), then this depreciation can be expressed diagrammatically by letting successive years become 30% shorter in the coordinate system for the cash flow curve, cf. Fig. 5.25 [14]. This means that we get a reasonable result when the areas under and over the curve are equally large.

The contribution to cash flow due to costs will normally have a long flat development corresponding to the early phases and the design activity. After this there is a sudden drop associated with investments in tools and equipment, the tying-up of materials, and

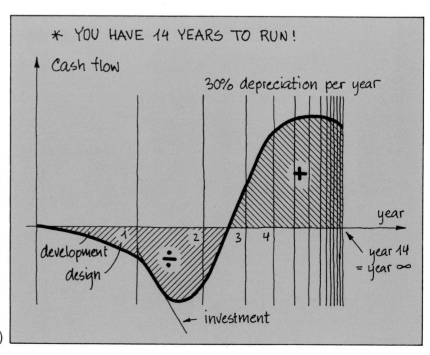

5.25 (after(14))

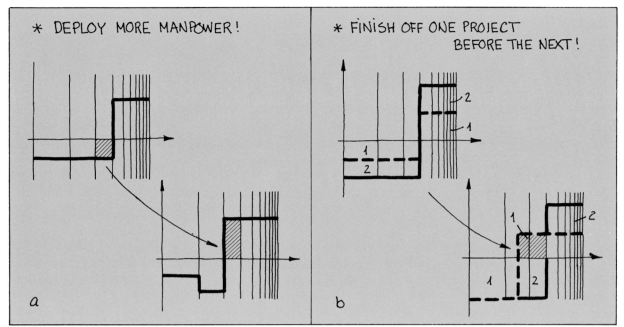

* DEPLOY MORE MANPOWER!

* FINISH OFF ONE PROJECT BEFORE THE NEXT!

a

b

5.26

so on. If we approximate this development by a series of rectangles, we can see the consequences of delays in Fig. 5.26a, where we, instead of allowing things to go on for longer, double our efforts in a final period: the area corresponding to the profit is larger than the extra costs. Thus it can normally pay to make a greater effort in order to avoid delays.

Figure 5.26b shows that it is an advantage to carry out projects one after another rather than in parallel, assuming that effort is proportional to results. Project 1 in the figure will give an early and therefore larger positive cash flow.

Cash flow curves should be drawn up early in a project, so they can act as a concretisation of the financial targets. But one will often experience a markedly worse result than estimated, because the project didn't proceed as planned. Figure 5.27 shows a typical result of this type, where the estimated cash flow curve produced at the time of the start of investment is compared with the one obtained in practice. The halved profit is caused partly by project costs rising by a few percent, a considerable part is loss due to delayed cash flow and a part which is twice as large is due to reduced sales (i.e. the reaction of the market to a product which is not optimal).

5.27

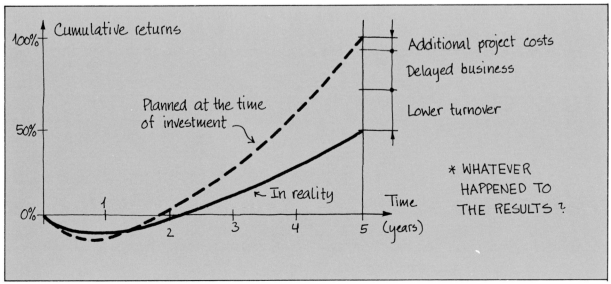

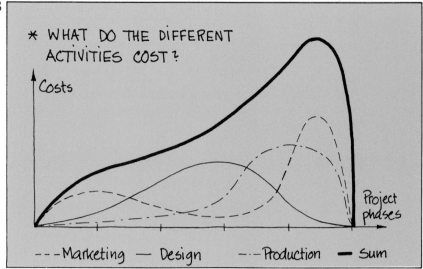

* WHAT DO THE DIFFERENT ACTIVITIES COST?

Costs

Project phases

---Marketing — Design ---Production — Sum

The costs associated with the three functional areas of marketing, design and production do not develop in the same way, Fig. 5.28 [27]. But it is more significant that there are costs due to their interplay, i.e. costs which are incurred due to the necessity of cooperation, the necessity of handing things over and communicating, and the necessity of correcting things due to poor interplay. Costs associated with this interplay may be considered as indirect development costs. Those who pay for them are only rarely those who cause them:

- Costs in the interplay between marketing and development, for example due to alterations in the product specifications, inadequate information or misunderstandings which are discovered too late, inadequate consideration of the relation to the customer, excessive consideration of technical points, incorrect or late information about volumes of sales, missed deadlines in relation to marketing, and so on.

- Costs in the interplay between design and production, for example due to design changes after the tools had been set up, changes due to lack of insight into production technique, or to out-dated ideas about it, changes due to failure to take details which are essential for the function of the product into account, costs due to quality control and test procedures in production, and so on.

- Costs in the interplay between production and marketing, for example due to missed deadlines in relation to marketing, insufficient or excessive growth of sales in relation to what has been planned, delivery times, as well as acknowledged servicing matters or questions of customer adaptation, and quality failures in relation to the specifications, etc.

It is difficult to make an exhaustive review of costs related to interplay; in what follows we shall focus mainly on some of the ways in which the way in which things hang together can be dealt with. In Chapter 8 we shall examine more closely some tools associated with interplay, which can be used to influence costs.

Dynamic problems in handling costs

In Chapter 1, we drew attention to the important law that governs product development (cf. Fig. 5.29). This was, that costs are allocated at an earlier stage and by other people than those who in fact appear to incur them.
This means that:

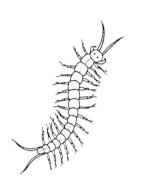

- The early phases of the project, in which everything is still in a state of flux, must be exploited by all interested parties; i.e. by people concerned with marketing, design and production, to the best interest of each of these areas.
- The early phases must likewise be exploited to the benefit of those areas which are the concern of everyone and thus perhaps of 'nobody', i.e. say indirect development and production costs, materials costs and several others.
- You need to be able to understand the relationship between decisions made in the early phases of a project (about the type of product, its principles, structure and design) and those costs which are a consequence of these decisions. This relationship is so complex that we often have no idea what we are up to!

5.29

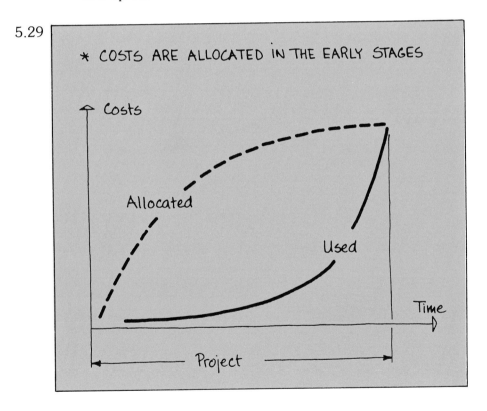

Thus the very nature of product development (Fig. 5.29), obliges us to make use of integration in the early phases. This means it is important that the different functional groupings within the company arrive rapidly and accurately at a common view of the task in hand – that there is no gap in their conception of the product[27].

A new point of view on costs

In the sensational report 'It's not so much what the Japanese do... it's what we don't do' by Ingersoll Engineers [28], it is pointed out that

5.30 (after(28))

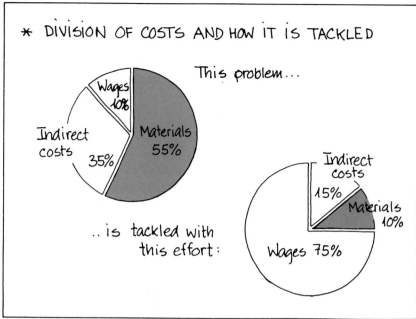

* DIVISION OF COSTS AND HOW IT IS TACKLED

This problem...

Wages 10%

Indirect costs 35%

Materials 55%

.. is tackled with this effort:

Indirect costs 15%

Materials 10%

Wages 75%

we use quite the wrong way of dealing with costs. Figure 5.30 shows a typical distribution of costs for an industrial product, together with the distribution of the effort used to reduce these costs. It is an outmoded idea to compete on wage costs – today the focus is on materials costs and indirect costs.

Most companies have entire production, design and accounts departments and a salaries department which conscientiously notes down the tiny percentage of direct wage costs. Are comparable resources expended on cost components which are four to six times larger – namely, materials? Of course not! [29]

"We lost sight of the cost problem – it moved some time ago" [28]

Control of materials costs lies with:

- The designer, who specifies the materials on the basis of the product specifications from the marketing department.
- The production engineer who specifies the size of the flow of materials.
- The purchasing engineer, who negotiates the price.
- The production department, which decides how things are to be produced, often on existing, inappropriate equipment.

Thus the responsibility for half of the costs of the product is divided among five groups of people, so that materials costs are in reality nobody's problem.

What then is an appropriate costs structure? It is a structure which affects the costs distribution of Fig. 5.31 within the large sectors, and which corresponds to the company's product- and production-strategic objectives, such as short development time, rapid production, small intermediate stores, late definition of variations, high product quality, and so on.

We get control of costs by considering and dealing with two streams as one. They are made up respectively of those costs associated with purchasing, production, sales and service, and those costs which are determined by the developed product's costs structure, costs in the development process itself, and costs which we recognise when the costs of the product are actually realised. This implies that we need to have the answers to a series of questions[29]:

104

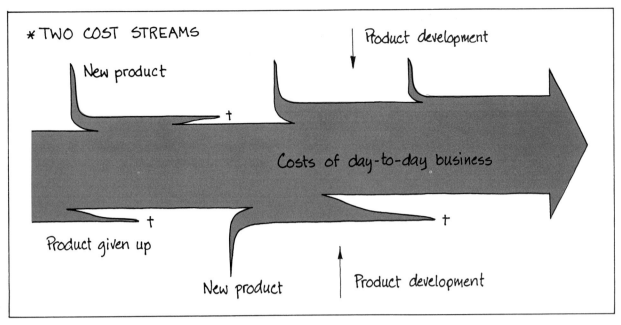

*TWO COST STREAMS

New product

Product development

Costs of day-to-day business

Product given up   †

New product   †   Product development

5.31

To the purchasing department:

- Which types of components are we going to develop and produce ourselves?
- From which types of market should we buy our components?
- Which type of supplier? What price policy?
- What prices, quantities, forms of supply, delivery quantities?
- What subsequent processing, checking and assembly is to take place?
- What quality of goods do we want to have supplied?
- What technologies should we buy information on?

To the production department:

- How do we avoid processing (e.g. by total forming processes), waste, scrap, assembly, internal transport...?
- How do we minimise storage amounts, delivery times, production times, changes...?

To the management:

- What is the costs structure for the product to be?
- What is due to materials, direct wages and indirect wages, respectively?
- What about capital investment: machines, buildings, the accumulation of knowledge?
- How much current account capital is required? What about production stores?
- What indicates that the project leader is aware of the significance of costs for business?

To the product development department:

- Which new products are to be developed? Using which production technology?
- What types, sizes and variations?
- Which market segments, and which users?
- What size production runs, what quality, what profitability?
- What are the competitors up to: purchasing, technology, principles?

To the central person, the designer:

- Are you aware what costs you determine?
- What pieces of information do you use to decide on the form of a product, so that it corresponds to the desired costs structure and the desired level of costs for the product?
- Where and how do you use this information in the course of the development process?
- How do you obtain data on other technologies than those which are already in use in your production processes or which are used by your subcontractors?

Conclusion

The costs structure of products at the development stage reflects and determines the costs structure in the day-to-day running of the company. If you want to affect costs, you have to manipulate the costs structure of your products while they are on the drawing board.

It is therefore necessary to have a usable model of costs within the company, especially for the many types of costs which can be affected by product development, in order to be able to deduce what to look for during the design process.

But we also need data which tells the designer what he is doing or what results he can expect when he does particular things. Thus we need to know unit and standard prices for machines, tools and operations, to know what the machines can do, to know what the costs associated with sales and service are, and so on.

It is the task of the management to ensure that 'somebody' keeps control of costs.

5.7 Making the most of the business!

A prerequisite for good results from product development is that everyone contributes to making the most of the business. We see that companies, for a variety of reasons, are divided up into independent departments with independent objectives. Thus there is a risk that employees do not have a sufficiently overall view to be able to contribute positively, and perhaps they don't even understand how to.

Making the most of the business involves three elements, as expressed in the model of Integrated Product Development as introduced in Chapter 3; namely the market, the product and its production. The task of the product development group is to create as much business as possible by determining the best possible solutions to problems in these three areas.

Seen from the designer's point of view, it is his job to create the greatest possible profitability for his product, i.e. the greatest possible difference between sales price and costs. Some working parameters of this difference lie in the areas of marketing and production (e.g. storage policy and pricing policy); the designer

• JUST A SCREW...

As an example of the relationship between early decisions and their resultant costs, let us look at a product with four screws, which hold down a cover plate and part of the 'insides' of the product. Screws give rise to costs:

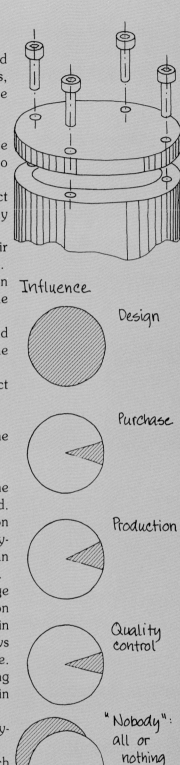

- in the design department, because they have to be designated by component numbers and their types have to be specified.
- for the purchasing department, who have to select manufacturers, make specifications, check the goods, pay the invoices, etc.
- for the stores, who have to unpack the screws, register their arrival, transport them, store them, find them again, etc.
- for the production department, who store them, put them in order, position them, screw them in and check the assembly.
- for the quality control department, who must check and maintain the quality of the screws, the quality of the assembly and the quality of the assembly equipment.
- for the sales department, who have to cope with a product where the screws contribute to the price.

Whose task is it to manage these costs? What influence do the different departments have?

- The design department can eliminate all costs. The designer decides whether screws are going to be used. Thus he also decides on which processes are to be used, on use of the production equipment, on tooling costs, quality-related problems and costs, and – more important than anything – he decides the cost structure of the product.
- The purchasing department handles a very large proportion of the costs, but has almost no influence on them. Successful purchasing can give discounts, while in unsuccessful cases it can affect the uniformity of the screws and thereby the cost of assembly and quality maintenance.
- The production department can rationalise the handling and mounting of the screws, and in this way attain a certain reduction in costs.
- Quality control can have some influence on those quality-related costs that are associated with production.
- The sales department has no influence on the way in which the screws contribute to costs. At best, the screws can be used as a sales argument ('easy to service'), at worst you can try to hide their existence.
- 'Nobody' has responsibility for the indirect production costs, of which many are related to the way in which the screws take part in the flow of materials, or as objects to be passed from one department to another. 'Nobody' has no influence, but 'we' have considerable influence, if we interfere in the designer's job!

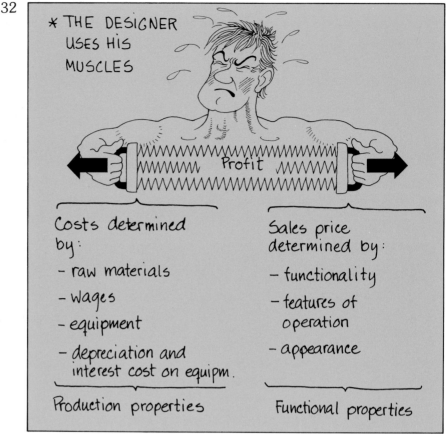

influences those parameters shown in Fig. 5.32. The parameters which affect the sales price are those properties which the customer values highly and is willing to pay for, i.e. performance, operational features, and appearance. We can consider all these as being functional properties of the product. Those parameters which affect the production price are raw materials, wages and investment in production equipment, together with leasing charges, amortisation and interest charges. We can denote all these production properties.

The designer's task is simultaneously to manipulate the structure, form and details of the product, so that the two 'prices', sales price and costs, are kept as far from one another as possible. Unfortunately, people in industry have a blind faith in the idea that these two prices are related through some kind of law of nature, which states that if you alter one of them then the other one must follow it in the same direction. Numerous examples show that this is not the case; the cure for this difficulty is better knowledge of the way things hang together.

The relationships already discussed imply that the designers have two clients for their results, namely, the marketing and production departments, and that the design department's role at some stage is played out (see Fig. 5.33). Remember to tell the design department this – perhaps they think that the technical aspects of the product are an objective in itself!

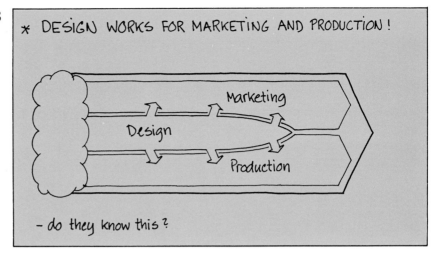

Figure 5.34 gives a conceptual picture of the course of a project, in which all three areas – market, product and production – are looked after. The picture has the form of a spiral. The project starts at the top and proceeds with an increasing degree of concretisation, through the areas associated with the market, the product and its production, where we focus in turn on these areas, introducing features of the required business. In all three areas there is a continual need for attention and for checking that an overall optimum is maintained, until the final proposal is reached at the bottom of the figure. A necessary requirement is that it is possible to work with very loose proposals and models at the start, and gradually make them more concrete.

Making the most of the business is the whole purpose of using a procedure divided into phases, together with a series of checkpoints, as we shall see in the next chapter. So at all times during the course of the project, the results obtained with respect to the market, the product and its production must be on a par with one another, and we must be able to put them together to evaluate the quality of the business and the probability of success.

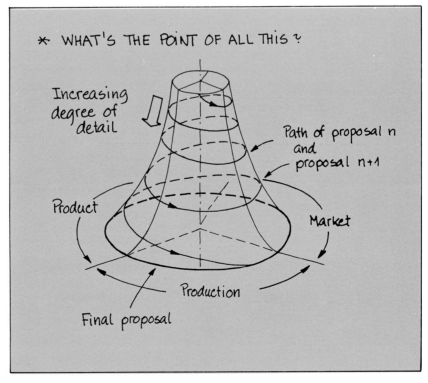

5.34

109

• THE DESIGNER'S DOUBLE LIFE...

As an example of the double role of the designer, who has to ensure a product which is both easy to sell and easy to produce, we show here a telephone handset in (a) an old and (b) a new design. The new design is based on the earphone and microphone being made out of drawn and folded flexprint, and the transducer out of piezoelectric PVDF film. The designer has to demonstrate two things: that the handset has the right performance (c), and that the fabrication of the flexprint can be carried out in a reasonable manner (d). Both these tasks have to be performed at the same time. (From an M.Sc. project carried out in collaboration with GN Telematic Ltd.) [43].

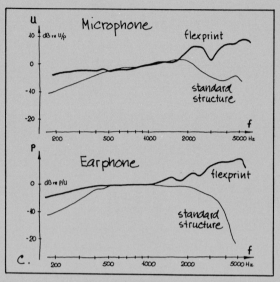

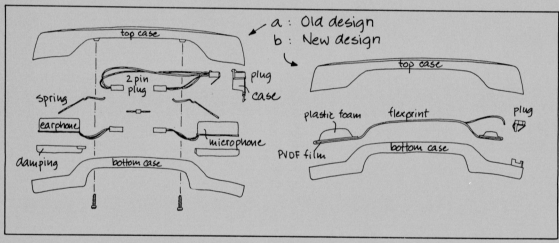

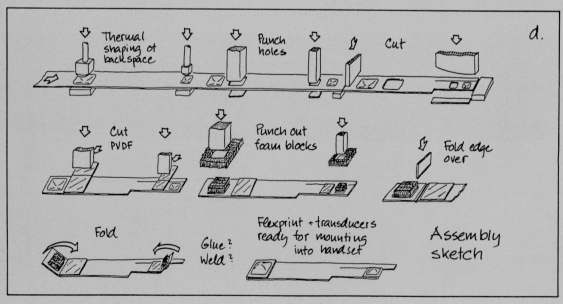

6

Creating the project

SUMMARY

In this chapter we shall look at the project leaders' tasks when setting up and starting a project. Different descriptive models of the development activity can be used in determining the type of the project and in determining what is to happen during the project.

Project management can be made easier by the insertion of keypoints, and adjustment of the project's objectives can take place by performing project design at the right keypoints.

At the start of the project it is necessary to fix a project strategy and to work out plans. The objectives for and assumptions behind the project can be maintained in a so-called Business Specification, and the desired product can be described in a Product Specification.

Now we are ready to start the project.

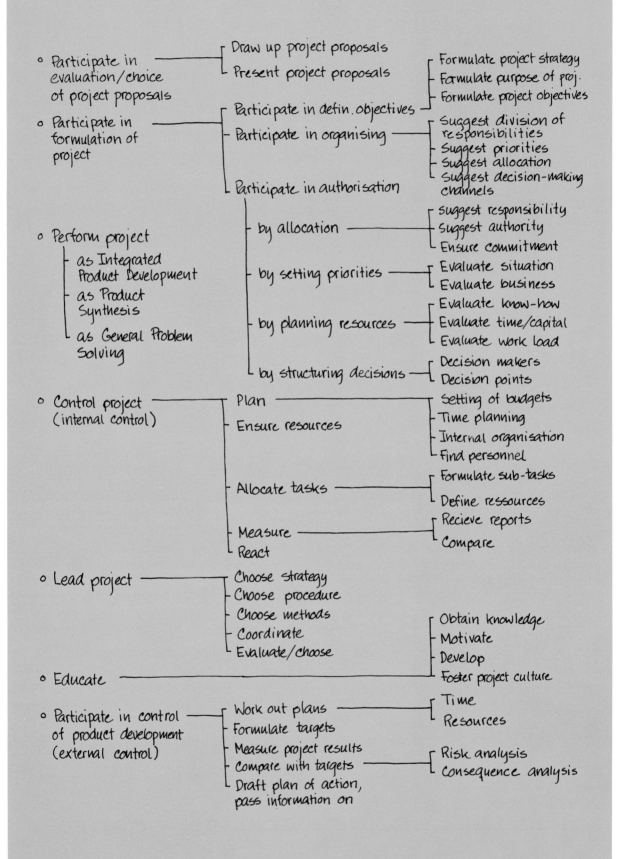

* THE PROJECT LEADER'S RESPONSIBILITIES IN PRODUCT DEVELOPMENT

o Participate in evaluation/choice of project proposals
— Draw up project proposals
— Present project proposals

o Participate in formulation of project
— Participate in defin. objectives
  — Formulate project strategy
  — Formulate purpose of proj.
  — Formulate project objectives
— Participate in organising
  — Suggest division of responsibilities
  — Suggest priorities
  — Suggest allocation
  — Suggest decision-making channels
— Participate in authorisation
  — by allocation
    — suggest responsibility
    — Suggest authority
    — Ensure commitment
  — by setting priorities
    — Evaluate situation
    — Evaluate business
  — by planning resources
    — Evaluate know-how
    — Evaluate time/capital
    — Evaluate work load
  — by structuring decisions
    — Decision makers
    — Decision points

o Perform project
— as Integrated Product Development
— as Product Synthesis
— as General Problem Solving

o Control project (internal control)
— Plan
  — Setting of budgets
  — Time planning
  — Internal organisation
  — Find personnel
— Ensure resources
— Allocate tasks
  — Formulate sub-tasks
  — Define ressources
— Measure
  — Recieve reports
  — Compare
— React

o Lead project
— Choose strategy
— Choose procedure
— Choose methods
— Coordinate
— Evaluate/choose

o Educate
— Obtain knowledge
— Motivate
— Develop
— Foster project culture

o Participate in control of product development (external control)
— Work out plans
  — Time
  — Resources
— Formulate targets
— Measure project results
— Compare with targets
  — Risk analysis
  — Consequence analysis
— Draft plan of action, pass information on

6.1 The tasks of the Project Leader

It is widely recognised that product development should be organised in the form of projects, and that projects must have a project leader. But it is not nearly so clear what is to be understood by a project leader, and thus what qualities you should look for in a project leader.

In this section we will look at the project leader's tasks and in this way indirectly indicate what he/she has to be able to do. In the following sections, we shall consider some of these tasks in more detail, and suggest some concrete tools which can be utilised.

The page opposite gives a general view of the tasks of the project leader. The diagram here is an elaboration of the diagram of management functions on page 58. The two diagrams hang together in the way that all the project leader's tasks are associated with tasks which are the responsibility of the executives or the management.

Notice that the boundary between the project leader's and the project group's tasks is fluid, just like the boundary between the management's and the project leader's tasks. The function diagram indicates some of the abilities which the project leader has to possess.

The principal task of the project leader is naturally to create business, i.e. to create the best possible combination of the market, the product and its production, so as to obtain good business results, but also in order to ensure that the company maintains a high internal efficiency. Important factors in the project are handling risks and contributing to technological development within the company.

Product development involves, as discussed previously, two types of ability: technical ability and project ability. A project will fail or run less efficiently if one or both these abilities are lacking. Project ability can be related to ability at solving problems, at product synthesis or at product development (see Section 6.2). But there are also some problems of interplay to solve – e.g. some characteristic situations which are reproduced here from a number of other sources and from our own experience:

- The range of activity is too narrow – the project group runs with blinkers on, design degrees of freedom are not exploited, alternatives are not considered.
- The group takes no chances – it lays constraints and norms on itself, it thinks and acts uniformly.
- The group is too bound to procedures – people don't discuss ways of doing things or results, each member works in his own way.
- Shortage of time and problems with setting priorities are the root of all evil – the group has discovered these scapegoats and exploits them.

A number of symptoms can give warning of these conditions: people keep on talking about good comradeship, extol the group, behave in an exaggeratedly secure manner, don't listen to outsiders, find external reasons for negative developments, use unity as an argument, make up problems in order to inflate their own job and give it more importance.

The project leader is an important person in the project; he needs to have abilities, skills and attitudes such as:

- Being a specialist within one or more of the specialised areas of the project.
- Being a leader, i.e. somebody who formulates tasks, motivates people, makes decisions or gives approval to decisions.
- Being a manager, i.e. one who makes plans and follows them up.
- Being an administrator, i.e. a channel for information to and from the project.
- Giving things structure, i.e. being the one who puts some structure into the project, has an overall view of things, divides it up into tasks, and so on.

But it is difficult to say to what extent he needs to be a specialist rather than a generalist, or to say which of the abilities listed above are the most essential.

Apart from possessing the functional qualities mentioned above, leaders must also take on a number of work roles, including:

- The role as standard-bearer, figurehead, official spokesman, example, motivator, team-maker.
- The role as solver of crises, conciliator, initiator.
- The role as project politician, the one who reveals what interests people have in the project, thinks strategically, keeps his eyes and ears open, has good connections.
- The role as negotiator, the one who can attain results.

In the following sections, we shall deal with some of the subjects which we believe belong to the project leader's area of activity and which make up his project professionalism. They underline his/her role as an integrator in several respects: between the management and the project, between the different areas of activity, between different projects, between short-term and long-term tasks.

6.2 The nature of projects

In the previous part of this book, we have mostly dealt with the framework of projects, their initiation, organisation, and so on – i.e. all those things outside the project which have to be in order for a project to be able to exist. We shall now look more closely at the project itself, at the way in which it is created and carried through, with the project leader bearing the responsibility, and working in collaboration with the project group.

What kind of pattern of actions and results is delineated as the project proceeds? What is the nature of the project and how should we take it into account in setting up and carrying out the project?

We shall here consider a project as that activity which leads us from

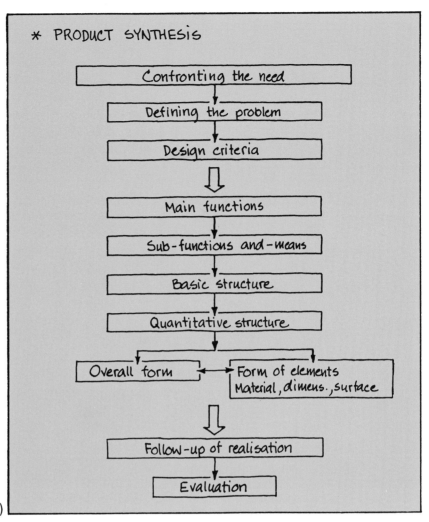

✳ PRODUCT SYNTHESIS

- Confronting the need
- Defining the problem
- Design criteria

⇩

- Main functions
- Sub-functions and -means
- Basic structure
- Quantitative structure
- Overall form → Form of elements Material, dimens., surface

⇩

- Follow-up of realisation
- Evaluation

6.1 (after(5))

a perceived business idea to established production and sales activities, which permit us to market the idea. Seen in relation to the pattern of integrated product development, a project therefore involves the three main phases of Development, Establishment and Realisation, as illustrated on page 25, into which other relevant phases are incorporated, depending on the level of ambition and the degree of novelty of the project. We have attempted to explain these relationships with the help of the elastic model shown on page 36.

A more detailed pattern turns up in a development project if we follow the product or machine as it gradually arises. The Product Synthesis model was presented on page 24. This model[5] (see Fig. 6.1) explains how one, starting with a formulation of the problem, can determine those functions which the product must offer, can find solutions in principle for each individual function, collect these solutions into a structure, determine the best structure and, on the basis of this, define which mechanical parts the product should consist of, whether these be machine parts, printed circuits, standard parts and so on, or parts which are produced specially for the machine in question.

The product synthesis model has the very important property that it handles or forms the framework around the so-called design degrees of freedom, as exemplified in Fig. 4.9. The design degrees of freedom is the tool which we can use to fix the principal extent, the 'length' of the development task. The number of subsystems,

115

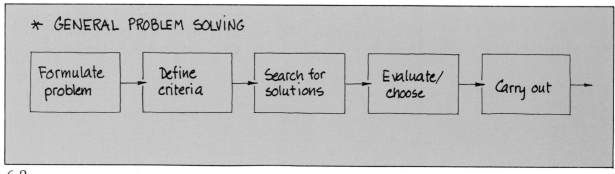

* GENERAL PROBLEM SOLVING

| Formulate problem | → | Define criteria | → | Search for solutions | → | Evaluate/ choose | → | Carry out | → |

6.2

i.e. the complexity of the product, depends strongly on the nature of the product, and gives the 'width'. Once you have fixed the length and width, then these together with the time give a measure of the project's 'volume'.

Note that integrated product development permits a total description of a project, while product synthesis only deals with the activity of engineering design, to which a number of elements have to be added in order for us to obtain the entire task which a design or development department has to deal with.

6.3

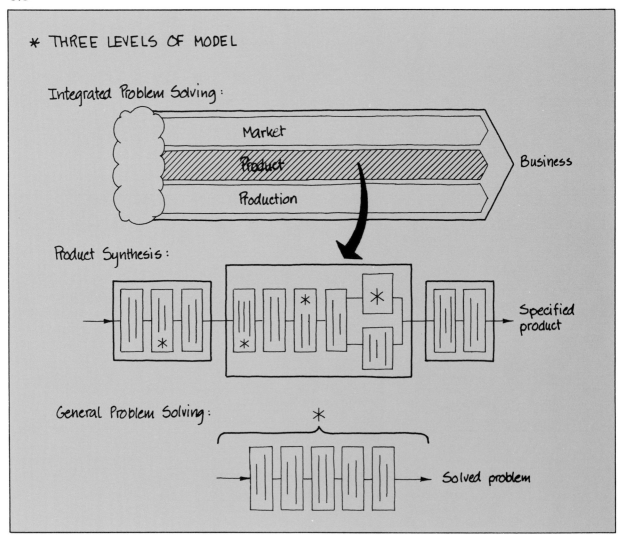

* THREE LEVELS OF MODEL

Integrated Problem Solving:

Market
Product
Production

Business

Product Synthesis:

Specified product

General Problem Solving:

Solved problem

116

A third description of development activities is General Problem Solving, which has also been introduced on page 24, see Fig. 6.2. Here we focus on the formulation of the problem, the definition of criteria, the search for solutions, evaluation and selection, and finally execution. This chain of activities must be considered as the least building block and will be included at many levels (e.g. in deciding on structure or form), and may make up a short or long sequence of events within a development project. Notice that general problem solving cannot be used to describe a complete project, but that we can use product synthesis to glue a number of these sequences together as a whole.

Figure 6.3 demonstrates the relationship between the three different ways of describing product development which we have presented here. We believe that these descriptions, the theories behind them, and the methods which are associated with the individual activities – and with the transition from activity to activity – reflect the nature of product development, and should therefore be mastered by the project leader.

Two things are missing in Fig. 6.3, namely an account of the way in which things progress within the marketing/sales and the production arrows. As far as the marketing arrow is concerned, quantities which characterise the use of the product and its sales parameters must be analogous to the design degrees of freedom, but we have not yet succeeded in finding theories covering this, nor in meeting anyone who dares to commit himself on this subject. With respect to the production arrow, we are on safer ground, as the characteristics of a process and for the development of the corresponding production equipment are the same as for the development of a product. When production machines are assembled into a production system, then characteristics such as the layout (structure) of the production system and the capabilities of the individual units come into focus. But, as we have said, this area is not at all well-defined.

6.3 Phases in integrated product development

The content of the phases of the Integrated Product Development model has been roughly described earlier. We think it is appropriate to present the phases in detail here, because in what follows we shall work with the model in a more detailed manner. Our starting point is the model shown in Fig. 6.4. We have not succeeded in finding an ideal project which can be described in all phases, and so we have chosen to describe the sequence of activities in a kaleidoscopic manner using many types and segments of projects. Thus our presentation is a 'longest sequence'.

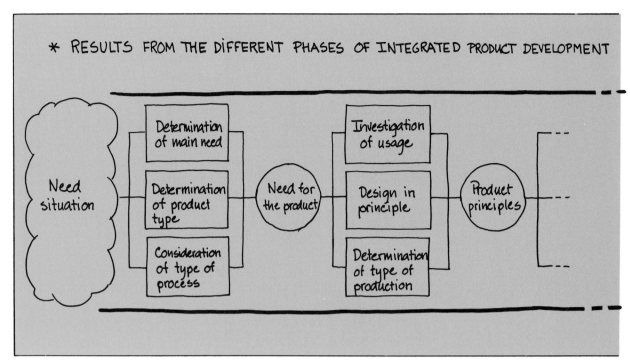

6.4

Phase 0 – Recognition of a need

The cloud in the Integrated Product Development model symbolises the need situation, i.e. the unacceptable or at any rate unsatisfactory situation in which we expect that our product can act to meet a need and can form the basis of a business opportunity. The problem or the task associated with this phase is to recognise the need, to discover the unsatisfactory situations.

Various search techniques can be used; e.g. one might conceptually look for more or less restricted starting points:

unrestricted
● Pollution?
  Oil: From underground tanks, from roads, from aquifers...
  Place: The Danish coast, Swedish lakes, Greenland...
● Transport by pipe? What can be transported? Where is this needed? Fresh water for Greek tourist islands?
  Waste water in Greenland?
● We can make plastic armoured concrete, what can it be used for?
  Canoes? Fire staircases? Bath tubs? Gateways?
● We supply hospitals with plastic articles. What do they need?
restricted
  Aprons, urine bags, tubes...

Apart from asking questions like this in order to create ideas, one can use: scenario techniques (changes in time, resources, attitudes); systematic searches, where particular characteristics (time, place, process, man/machine relations, control, etc.) are varied; system identification techniques, and so on. Or one can basically work 'with one's antennae out', ask, discuss, read journals, visit exhibitions, and so on.

The output from this phase can be a series of diffuse perceptions of

118

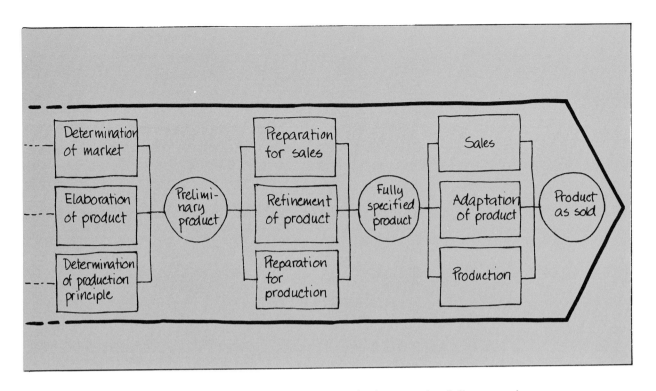

needs, which are worked on in the following phase.

Phase 1 – Investigation of need

'Need in principle' means here a need which could potentially be met, but where we have not taken market mechanisms into account. The establishment of the fact that there are fresh-water lakes in Sweden which suffer from oil pollution can lead to the idea that one might develop an 'oil remover', but this is no guarantee of the fact that anyone actually wants to buy or use an oil remover. Needs in principle are thus of the form:

- Tennis courts + rain = 'water remover'
- Waste water + Greenland = 'heated sewage pipes'

By 'product type' is meant the product as defined through the way in which it is used, i.e. defined by the way in which it meets the need. The product type can be determined on several levels, for example:

- transmission
- mechanical transmission
- mechanical transmission based on friction
- mechanical transmission based on friction between plastic and steel, in particular between belts and pulleys

Thus the development project can have as its starting point a more or less well-defined product type.

By 'process type' we are thinking of a very general classification related to the establishment of the product: should we buy, produce ourselves, lease, hire, and so on. For example, we might consider building a pedestrian bridge over a motorway on the spot or producing it in a workshop, while a pipe for waste water could be produced by winding, extrusion, assembly from shells, casting in place, and so on.

119

In this phase, the use of need trees, listing of attributes, input/output modelling, process generation and resource analysis are examples of important methods.

Output from this phase is a specified need expressed in terms of product type, need in principle and process type.

In addition to this, we need to work out the project goals (a specification of the idea, objectives, basis, requirements and criteria for the project, so that the goodness of the project, expressed in terms of its business possibilities, can be measured), the product goals (formulation of the problem, requirements and criteria, so that the quality of the product can be measured), and a time and resource plan, see Sections 6.6 and 6.7.

Phase 2 – Product principle

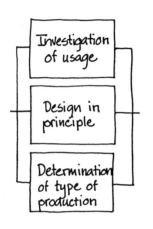

The aim of the user investigation is to clarify and specify the situation in which the product will be used (including the division of activities between man and machine) at all stages of the lifetime of the product: during establishment, while in service, during maintenance and during destruction. A number of alternative ways of meeting the need must be considered. Important methods are analysis of the competition, projection, trends analysis, resource evaluation, and consequence analysis.

In the product principle design phase, consideration of the process and functions leads to the determination of the functions of the product, after which solutions to individual sub-problems or sub-functions are sought and combined into a total solution. In this phase it is important to get an overall view of the solution space, which includes determining what stage the competitors have reached, from a technical point of view.

If new principles or operations are to be brought into use, it may be necessary to demonstrate, by means of models or experiments, that these principles can in fact be used.

The principles which are selected and the ways in which the product can be structured enable us to determine the types of production, i.e. to clarify questions such as: Is it a question of electronic or mechanical production? Do we need new forms of production? Does the product have a structure which will give a sensible production layout?

The output from the product principle phase is a product specified with respect to the principles involved, i.e. all principles are fixed, the structure of the product has been determined, and keypoints with respect to its function have been clarified. In addition, the situation in which it will be used has been specified, the relationship to the competition as regards the function and use of competing products has been determined, and the nature of the production has been fixed. A revised timetable and an updated set of targets are worked out.

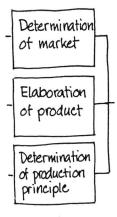

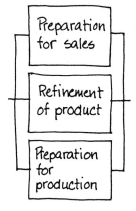

Phase 3 – Product design

Market investigation involves the clarification of product specifications, the determination of sales channels, pricing policy, launching, and so on. A very important point here is the estimation of the volume of sales, amongst other things, based on a price estimate from the preliminary product design phase.

Preliminary product design involves determining the total form, fixing details and design, and preliminary considerations of how the product is to be produced. The functional suitability of the product is demonstrated by dimensioning and prototype experiments.

The principles of production can be determined at the same time as the process and assembly are considered in a preliminary way, especially if the realisation of the product depends on using special production techniques. If this is the case, then the principles to be used in the special equipment are fixed in this phase.

The output from this phase is a completely specified product, except for the final questions of how to produce it. The product's function has been demonstrated and a rather certain estimate of costs has been made.

The market has been shown to be a likely one, and its magnitude has been estimated. The likelihood of a successful production has been established. A revised timetable and an updated set of targets are worked out.

Phase 4 – Production preparation

In the phase of preparation for sales, sales predictions, detailed plans, brochures and advertisements are worked out. Modification for manufacture involves a detailed determination of the components of the product based on the already defined production and assembly procedure.

As an important result of this, the workshop costs associated with the product can be worked out.

In the phase of preparation for production, the production system is defined in detail and set up, materials are purchased and the product is incorporated in the company's planning and management systems. By means of a pre-production run, the suitability of the product for production is demonstrated.

The output from this phase is a product specified for production at a well-determined price, sales have been planned and sales channels established, the production system has been set up and the suitability of the product for production has been proven.

A revised timetable and set of targets are worked out.

Phase 5 – Execution

Sales activity involves initiating sales, maintenance of the sales organisation, management, feedback, fixing pricing policy, and so on.

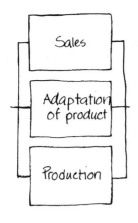

Product adaptation involves, in the short term, adjustment of the product's function and production, and in the long term, the accumulation of marketing feedback, updating of the product and changes to the product.

Production involves all those activities which ensure the establishment and maintenance of running production. The output from this phase is a product which can be sold as a proof of the product's saleability, and its functional suitability seen from the customer's point of view.

As a result of this, a total project evaluation and re-costing should be carried out.

Many aspects of a project have been omitted in this presentation – amongst other things we may mention financial control, decisions at keypoints, etc. – but we hope that the presentation gives a reasonably complete picture of the idea and contents of the Integrated Product Development model.

6.4 Procedures

Many companies make use of procedures, i.e. outline plans, as starting points for organising and carrying out development projects. In this section we shall look more closely at the use of procedures: What do we mean? What sorts are there? What does a procedure for integrated product development look like?

What is a procedure?

As stated above, a procedure is a model or pattern in outline form, which can be used when working out a project plan. As a rule, a procedure has several purposes:

- It must put structure into the development task, so that this proceeds efficiently and safely – in other words, you shouldn't be able to gain anything by deviating from the procedure.
- It must divide the course of the project up into a number of phases with a clear result coming out at the end of each phase.
- It must support the management in formulating, controlling and following up development projects.
- It must support the project leader in planning and following up a development project.
- It must give all those involved in the project a common frame of reference and an understanding of what is supposed to happen and, in broad terms, who is supposed to do what.
- It is helpful in the building up and transfer of experience with respect to product development, as regards methods, courses of action, norms, use of resources, mistakes, and so on.

The form of a procedure depends on a number of factors:

- How repetitive the development task is. If the projects are very much the same, e.g. in the form of campaigns or fixed changes of model, then there is every good reason to place development work into a fixed framework. If, on the other hand, you want a suitable spectrum of projects involving both modifications and new developments, then you should be more cautious about including at any rate the early phases within the procedure.

- The extent to which the organisation is adapted to product development. If product development is a rare and strange event, then this can be a reason for using well-formulated procedures and a well-defined project organisation. If, on the other hand, you have a typical development organisation, then procedure and organisation will almost blend together. In borderline cases, the course of the project will exactly follow the way in which paperwork moves round in the organisation.

- The extent to which development activities are well-established, or how well-known they are, at least as far as key persons are concerned. The innovative creator of an enterprise has no need for procedures, but when the company grows and groups of employees, some of whom are newly trained, are selected for the task, then it becomes necessary to have some form of procedure.

- The extent to which a project culture has been developed. A whole range of things can be taken care of by agreement among colleagues, standards of performance, creative interplay and so on, rather than by decrees arising from a procedure.

How does a procedure arise? Most probably by the crystallisation of experience accumulated from planning a series of well-carried-out projects, although it can happen on the basis of more or less valid theories. Let us now look more closely at some of the types of procedure which are available.

Common procedures

In the literature, many different procedures for product development are suggested, and this gives rise to the first question: Is there a universal way of going about product development? On the basis of what we have previously stated in this book, the immediate answer is 'No'.

If by a procedure we mean a sequence of activities which is to be followed in all development projects and which will get to be included in every project plan, then the answer is 'No'. If, on the other hand, we mean a framework or skeleton, of which larger or smaller portions can be incorporated in a current project and where the framework can be filled out with just those ways of doing things which are appropriate, then the answer is much closer to being 'Yes'.

In the literature, there are many bad examples and we shall refrain from rehearsing them here. They are bad or wrong on one or more points:

● DEVELOPMENT PROCEDURES AT VOLVO

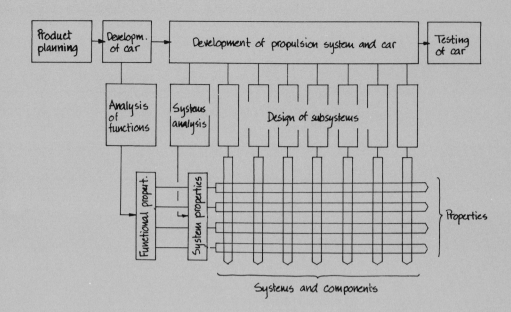

As an example of a product development procedure, we present here an exceptionally interesting method used at Volvo Cars AB. The product is considered partly as a carrier of functions and properties, partly as a whole system. During the development process, subsystems such as the chassis, transmission, motor, and so on, are created by separate groups, and important properties such as ergonomics, ease of operation, climatic comfort, and safety are dealt with by separate groups, as shown in the figure. The two results, which we might call 'the technology' and 'the dream', are in this way created by experts in the individual areas.

- They focus on an 'idea' as the essential thing. Recognition of the facts that product development originates from a need, that a large number of possibilities are consumed on the way, that three results are needed to create business (a market idea, a product idea and a production idea), and that as many ideas are required to make a product as are demanded by the product's complexity and the design degrees of freedom – none of these things can be included in the limited concept of an 'idea'.

- They mix up the three levels of description shown in Fig. 6.3. You find procedures which involve general problem solving, garnished with elements of a product development sequence, but you rarely see attempts to incorporate the nature of the product, as it is reflected in product synthesis. How can one avoid talking about function, principles, structure and form in a procedure for the creation of mechanical products?

- They include particular choices of ways to carry out development work – for example, a particular strategy which says that the product's design should be fixed at an early stage, or particular tactical techniques, such as investigation of patents, production of working models, market analysis by means of prototypes and so on. These methods will perhaps be suitable in one situation and not at all in another. If they are incorporated in the procedure, then they will perhaps be included in the plan for the project without anyone consciously thinking about what the best choice of technique might in fact be. At the operational level we also find a selection of particular techniques, e.g. brainstorming or morphological methods as techniques for creating ideas, or use of weighted criteria as an evaluation method. There are others!

Thus there is every reason for being sceptical about procedures, and also to be sceptical about other companies' procedures: Would they suit my company? Do we have the same type of products? The same degree of renewal? The same frequency of renewal? Should we choose the same strategy? Methods?

At Danfoss A/S, the procedure for realisation activities is a detailed one, while it is only sketched for the earlier phases. The procedure shows, in the form of a detailed network, how the company's many specialised departments are brought in to solve sub-tasks: approval, patents, packaging, testing and so on. At B&O A/S, the procedure includes at an early stage the preparation of a design model, which subsequently acts as the specification framework for the design task. Volvo Personvagnar AB have adopted an interesting development philosophy as the basis of their procedure and organisation (see opposite).

Which procedures are used outside Denmark? In the USA, according to Booz, Allen & Hamilton, people tend more and more to take a formal view of the development process, but there is nothing that indicates the use of procedures. In the American literature, only the level of general problem solving is treated, and our experience from a study tour [25] is that companies usually choose the 'right man' as project leader, and that all problems are then considered solved. The 'right man' means, amongst other things, the one who has tried it before.

In Japan, written procedures and theories are not used either. We have found no signs of these in material from conferences or in companies which we have visited [24]. Instead, they painstakingly provide for the training of project workers according to a rota system, so that after 7-8 years of experience they can take part in a project. The project is led by an experienced man who chooses his own way of doing things.

In the UK, procedures oriented towards problem solving are used, but only by a few companies. There is a heavy demarcation line between design and production, and thus a considerable need for integration. In Germany, large companies make use of formal procedures for product planning and engineering design, but follow a sequential basic pattern.

An important question with respect to specific company procedures is the extent to which certain parts of the sequence of activities should be specialised or institutionalised. Many arguments indicate that it is an advantage to collect expertise about, for example, business searches, tool design, patent matters, design of special subsystems, the preparation of drawings, the choice of tolerances for components, the registration of drawings and so on in small departments, but this can have serious consequences for the dynamics of development activities. This can easily lead to the questions: Does the project merely go past the individual participants – or does he/she run with the project? (Fig. 6.5).

It is interesting at this stage to consider what the procedure is supposed to support or preserve within the company. The significance of decisions taken during the course of a project can be

6.5

* DOES THE PROJECT MERELY GO PAST THE INDIVIDUAL PARTICIPANTS ?

– OR DO THE PARTICIPANTS RUN WITH THE PROJECT ?

6.6

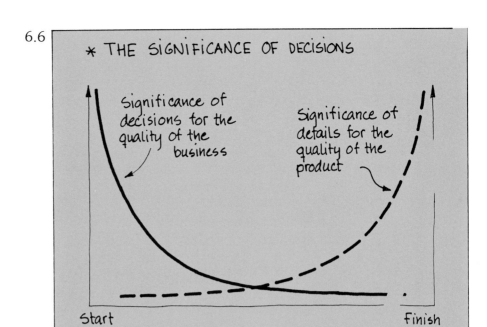

illustrated as shown in Fig. 6.6. One of the curves shows how the significance of decisions for the quality of business falls rapidly as the project proceeds ('external efficiency'), and the other shows how the significance of decisions for the details of the product, and thus for the product's quality ('internal efficiency'), rises towards the end of the project. A large company, with all its resources and specialists, can incorporate more quality into the details, whereas it

6.7

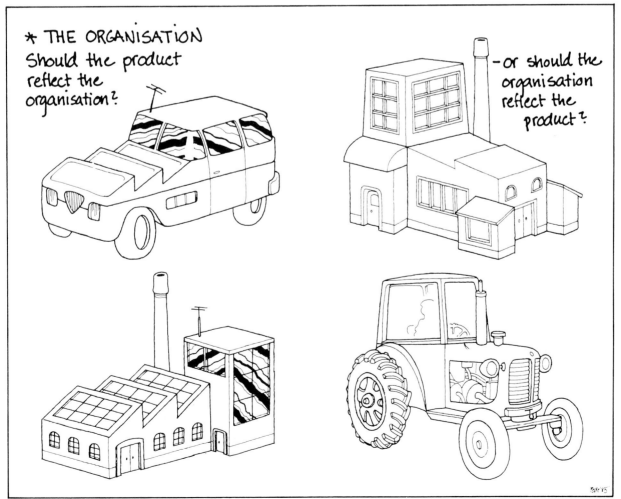

to a certain extent is prevented from incorporating quality into its business, due to having a company structure which reflects the structure of the existing products. The company has started to look like its product (Fig. 6.7).

Keypoints

Let us return to the question of the purpose of a procedure and look at an important aspect of procedures, which is their ability to make projects quantifiable and controllable for the management. The management needs to evaluate the feasibility of the project and its results at various critical points – at the transition from preliminary investigations to formulation of the project, at the time of investment in production equipment and marketing, and at the time when the product is launched.

And both the management and the project need a sequence which is divided up in order to:

- promote clarity
- ease planning
- ease control
- minimise risk
- optimise business

It is therefore a good idea to insert keypoints into the project – i.e. deadlines at which the business operation, the risk and the project are evaluated and possibly adjusted.

The creation of results or the degree of definition or the rate at which we become more knowledgeable do not rise constantly as a function of time, but exhibit some static periods (see Fig. 6.8). It is a good idea to place the keypoints in these periods. As we know, three results should be presented at each keypoint – namely results related to the market, the product and its production. Thus the keypoints should be placed where we can note a marked rise in our understanding of all three areas.

The phase transition in the Integrated Product Development model has exactly this property, as can be seen from the detailed presentation of Section 6.3. Thus a keypoint plan can appear as

6.8

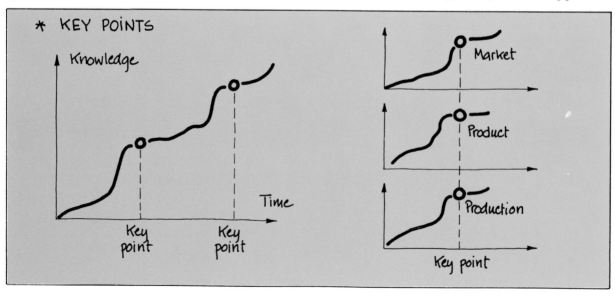

128

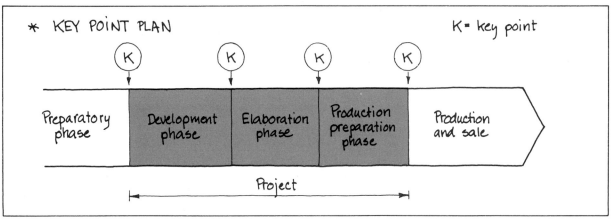

6.9

shown in Fig. 6.9, although one could choose to include more or less points, depending on the nature of the product development. The keypoint plan shown here is characterised by the actual project starting when the business search has finished and the project preparations have been terminated (critical points have been investigated). After this we want to keep control of the result of the development phase (i.e. definition of the usage and principles of the product), of the elaboration phase (which for example terminates with the evaluation of experimental models) and of production preparation (which for example terminates with the evaluation of a pre-production run). This keypoint plan is well-suited to companies which at least from time to time seek after new technology and new uses for known products, markets or production systems.

The early activities – i.e. recognition of the situation, business search and preparatory activity – can also be made project-oriented, but it should be pointed out that these activities cannot be 'named' in the same way as product projects (Fig. 6.9), and that this can lead to multi-stringed sequences of activities, as shown in Fig. 6.10.

6.10

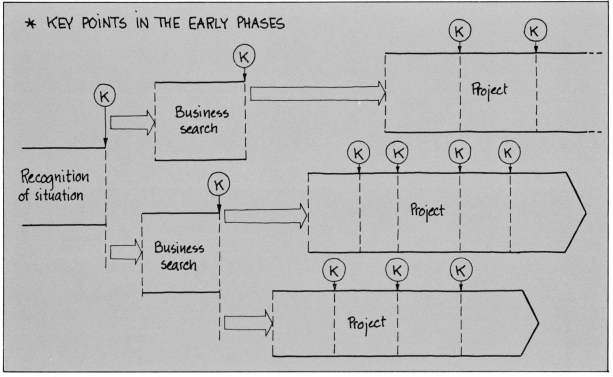

If product development involves or to any great extent is based on research within the company, then this must be taken specially into account. Many types of research cannot stand being made business- or product-oriented, and so research should not be included into the product development sequence. We shall return to this point in Chapter 9.

In the keypoints, the following things can (or ought to) take place:

- The status of the project is measured, decisions which have been taken are evaluated.
- The result is evaluated in relation to the project objective (BS, see Section 6.6).
- The result is evaluated in relation to the company's other activities.
- The consequences of stopping and of continuing the project are evaluated.
- The decision is made as to whether the project is to be continued (or stopped).
- The project is designed (see below).
- The status of the project and the decisions made are noted.

Some of these points are considered in more detail in the remainder of this chapter.

Project design

We have claimed earlier that the project organisation ideally speaking should be altered during the course of the project in order to adapt it to the demands which are made at the different stages. In addition, external conditions (increased demand, threats from the competition) or project results (a world-shattering product, the need to catch up on a deadline which has been missed, division of a project into parts because it has become too large, and so on) can naturally lead to a need for adjustments in the organisation or targets.

6.11

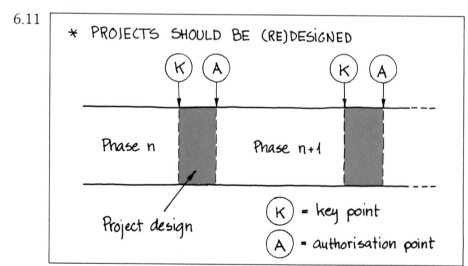

In principle, a project should be (re-)designed at each keypoint

Project design is an activity which the project leaders and the top management should initiate, not as an item during a keypoint meeting, but as a task which is to be terminated at an authorisation meeting (see Fig. 6.11). The project design should involve:

- Formulating/adjusting the project targets.
- Formulating/making more precise/giving more details of the targets of the next phase.
- Determining the resource requirements of the project.
- Making the timetable for the remainder of the project.
- Evaluating the risk associated with the project.
- If necessary, inserting checkpoints in the following phase.

By checkpoints is meant additional status points which are inserted in order to focus on particular aspects of the project. They can be related to time (deadlines) or to results.

Conclusion

The concept of a procedure, seen in relation to integrated product development, is subject to the extremely important observation that:

> Product development must take place on the terms of the project, not on the terms of the organisation

Projects are different, because the conditions – the way in which the business is dealt with – either taken individually or all together, alter from project to project (Section 6.6). Our attitude to procedures can thus be summarised in the following statement:

> A procedure only needs to consist of a keypoint plan

It is the task of the project leader to organise the project within this open and free framework. If the company finds it useful, then checklists or handbooks can be worked out for certain activities between keypoints, but basically there should be a considerable amount of freedom to choose a way of going about things within a very secure and well-defined pattern of keypoints.

6.5 Project strategy

The concept of strategy covers methods used to realise long-term objectives, and can be applied at several levels within a company. We saw in Section 4.2 how the company's strategies, amongst other things for product development, were used to choose among alternative business ideas. Thus the company's top management selects particular projects as a means for the realisation of the company's business targets.

The top-level product planning or product development strategy should be well-structured and should ensure a volume of product development which gives the right products at the right moments. It should be based on a suitable technical effort, in accordance with a well-organised, progressive programme of work, and with a suitable degree of preparedness in relation to the elements of risk which are always associated with innovation [30].

If we consider the company's individual areas of activity, such as the three which are of particular interest in connection with product development (marketing, development/design and production), then strategies are also available for these areas:

- Marketing strategies, which determine which marketing resources are to be used on the individual market segments, which markets one wants to win with what products, how to accumulate greater knowledge for application to certain markets, whether we go in for offering service for large products, whether the range of certain types of product is to be increased, etc.
- Design strategies, which determine which principles one should go in for, what ranges of sizes a product is to be produced in, what family relationship there is to be between two product types, and so on.
- Production strategies, which determine quality targets, automatisation targets, whether to go in for particular processes and materials, the accumulation of know-how, make-or-buy objectives, purchasing strategy (e.g. spreading a job over several subcontractors), the strategy for delivery capability, and so on.

These strategies must be implemented through the development projects. The company can only change course as a result of development projects changing its business base and cost structure. The relationship between the strategies belonging to the three functional areas and their implementation in a development project is not a simple one, but takes on the form shown in Fig. 6.12. It is particularly important to stress that the area of development/design contributes very strongly to the realisation of the marketing and production strategies. Or, to put it the other way round: it is within the areas of design and development that marketing and production conditions can be influenced decisively.

If each of these areas wants to take its own development strategies into account, for example by ensuring a shorter delivery time or by ensuring the introduction of automatic assembly, then they must

6.13

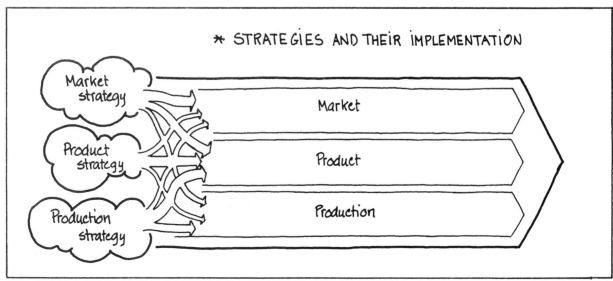

132

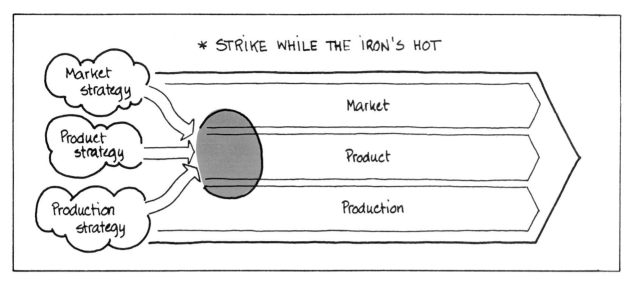

6.12

'buy' influence by offering some manpower, so that the early phases become genuinely integrated product development. It is in the product area that the greatest influence can be exerted, but it must happen while 'things are in a state of flux' (Fig. 6.13).

Among the strategic targets which today are singled out at being important for the production area are reduction of wage-dependent costs, the attainment of great reproducibility (e.g. by increased automatisation) and staking a lot on the effect of experience. Experience shows that these can radically be influenced as a result of design activities.

How methods which help to realise the company's marketing, product and production strategies are incorporated during the planning of the project will be discussed in Section 6.8.

When the decision to establish a project has been taken, the project leader is in the situation where he must select the necessary project strategies – i.e. those strategies for carrying out the project which will ensure that the desired results are achieved. The primary project target is to establish and make the most of some business operation, but this involves a number of sub-targets such as quality, finance, efficiency of development, minimisation of risk, and so on. Some examples are:

- New products can be associated with so much uncertainty, that one settles for authorising the early phases of the design process, so that the management does not commit itself to the whole project. This is a project strategy dictated by the management.

- The project leader can choose to make a frontal attack on the project. Or he can choose to find the points and ways of doing things about which he is certain, and those where there is uncertainty, and to fit them together, while all the time attaching most importance to the uncertain ones (Fig. 6.14).

- The project leader can attach particular importance to integration, and can attempt throughout the organisation of the project to avoid sequential procedures and transfers of

133

6.14

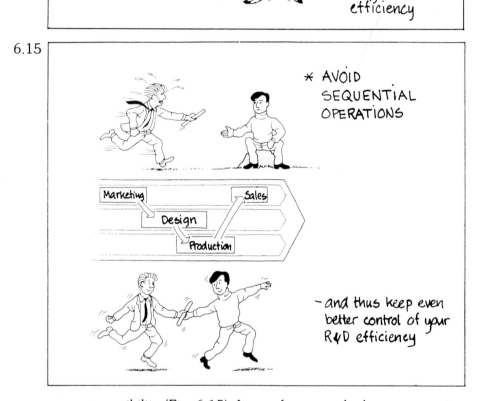

* ALLOW FOR THE CRITICAL POINTS

— and thus maintain control of your R&D efficiency

6.15

* AVOID SEQUENTIAL OPERATIONS

Marketing

Design

Production

Sales

— and thus keep even better control of your R&D efficiency

responsibility (Fig. 6.15). It can, for example, be necessary to use measures of this type if the project is to be particularly short, so that one chooses to establish provisional production equipment. In this case, the department responsible for production technique must take part in the structuring of the product and the detailing of the design, so that the production equipment can be defined as early as possible.

- The project leader can choose to sketch the outline of the succeeding development project, in order to put the wealth of disturbing elements in the form of changed marketing requirements – 'we just have to...', and so on – off to the next project. This is important for the efficiency of the project, and in order to keep deadlines (Fig. 6.16).

- The project leader can aim for an early clarification of the saleability of the product, in order to remove part of the risk

6.16

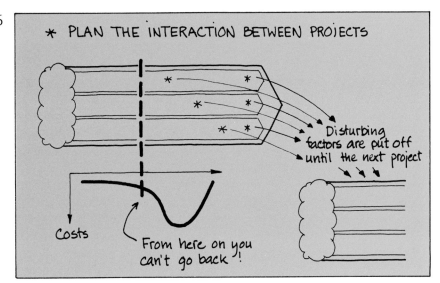

* PLAN THE INTERACTION BETWEEN PROJECTS

Disturbing factors are put off until the next project

Costs

From here on you can't go back!

involved in investment in production equipment and in the marketing process itself. This can, for example, take place by discussing prototypes or design models with selected customers. Or he/she can aim for early clarification of the statistical uncertainties related to the function of the product, and can therefore organise a pre-production run in order to eliminate part of the risk involved in marketing. The products should preferably not come back from the customers like homing pigeons!

A prerequisite for being able to choose a strategy is that you are aware of your own situation. The project leader is not just somebody assigned to that particular slot – he must also have considerable knowledge of the company's affairs, and be able to see all the economic aspects of the project and its significance to the company.

For example, it is important for the project leader to recognise what phase of development the product is in, and what pitfalls lie in each phase. At Danfoss [14], the various phases and the associated risk factors are described in the following way:

- The introductory phase is characterised by considerable uncertainty with respect to market requirements and future development, and therefore demands that the company demonstrates a marked willingness to take risks. The market is characterised by a few product types with very broad specifications. Price plays only a small role. The products have many teething troubles, and efforts are made to master the technology of the product. In this phase, considerable backing from management is required if success is to be ensured. The reasons for failure are quite characteristic: too little effort, too little backing, excessively close control, absence of a standard-bearer.

- The growth phase is characterised by intensive study of market requirements and applications. The market begins to show signs of dividing up into segments, so that a greater breadth appears in the programme. Increase in production numbers leads to a need for new production processes. Models change rapidly, so considerable development effort is

• VARIABLE COSTS OR THE EFFECT OF EXPERIENCE?

One of the most important measures of control is the proportion of variable costs. Many projects fail because this cost target is not reached. But it is reasonable to ask whether the first products to be sold need to reach the cost target. Wouldn't it be better if it were reasonably likely that the products could be made much cheaper as a result of experience?

At B&O they use a strategy in which they aim to get the products on to the market as quickly as possible, so the pattern of sales can be discovered. After this, substantial rationalisation takes place in the production technique, based on considerable, but carefully planned, changes in the design of the cost-heavy electromechanical construction.

The pictures here show the series of Beogram gramophones with tangential arms. (A) shows the Beogram 4000 model, whose internal construction in the first version (B) is based on the electronics being divided among several printed circuit boards, and the use of an internal chassis of zinc. Type 4002 (C) is the result of an evaluation which led to a series of simplifications.

Rationalisation of the product continues in the model Beogram 8000 (D), in which the internal chassis (E) is produced as an outsert moulding, and the drive is altered from a belt drive to a linear motor. The structure of the product is altered so that a hinged top carries the electronic components. In (F) we see the model Beogram 6000, which gives a further reduction in costs, and where return has been made to use of a dc motor and a belt drive.

A

B

C

D

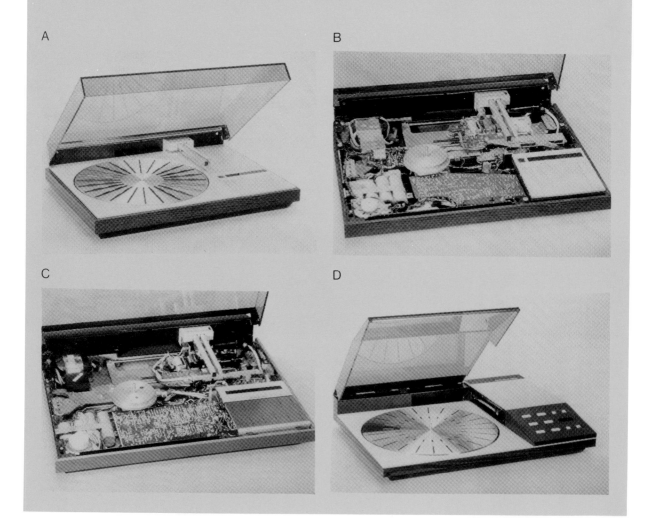

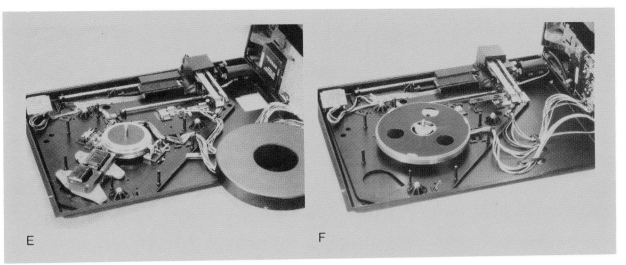

E   F

required. The reasons for failure are insufficient knowledge of applications, too little effort put into product development, and too slow a build-up of the marketing organisation.

- The maturity phase is characterised by the battle for market shares – products must be as good as possible with respect to costs and quality, and all changes in the products or their marketing must be carefully thought through. There are very few changes of model, and tight control with very little taking of risks is essential. The market is strongly divided into segments. Development activity is concentrated on updating, on development of new market segments, and possibly on changing technology. The pitfalls at this stage are choosing the wrong technology, an incorrect marketing strategy, staking too little on preserving one's competitive edge, and having too low a quality of product.

6.6 The Business Specification (BS)

In this section we present a project management tool – the so-called Business Specification (BS) – which is used as a contract between the management and the project group (especially the project leader) to specify the targets which are to be reached.

A project cannot be run in accordance with the intentions of the management if things are managed solely on the basis of the specifications for the product. The objective of a project is normally to create some business for the company, i.e. the best possible combination of market, product and production. We saw in the previous section that the project must be executed in accordance with the strategies agreed for the relevant areas of activity. Thus, in order to ensure the development of these areas, the BS must relate the project to commercial considerations and to the investigations which preceded the project, and it must unify the management's task of directing the project. The BS marks the design of the project, see Section 6.4, and contains:

- The commercial targets, which the project must lead to, and which consist of results associated with marketing, product and production.

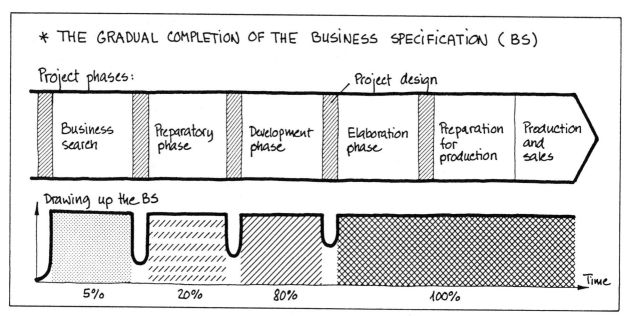

* THE GRADUAL COMPLETION OF THE BUSINESS SPECIFICATION (BS)

Project phases:

Project design

| Business search | Preparatory phase | Development phase | Elaboration phase | Preparation for production | Production and sales |

Drawing up the BS

5%   20%   80%   100%   Time

6.17

- The project's financial targets and organisation, i.e. the main financial results which are looked for, and the way in which the project will be financed.
- The conditions under which the project will run, i.e. its staffing, its limits, its resources, the way in which responsibility is organised, and the way in which it is to progress.

The BS is worked out by the management in collaboration with the project leader, at the time when the project is set up. The BS can also be used successfully in the early phases – i.e. for management of phases of business search, preparatory investigation and development. The BS is adjusted and expanded in step with improvements in knowledge, increases in definition, and decisions, and it reaches its final complete form at the launching of the design project (see Fig. 6.17). Note that the BS is complete in all phases, but only contains a few details in the early phases.

The BS is utilised in two management activities. One of these is product planning, i.e. the management-related control of all product development activities, including business searches, initiation, start, follow-up and exploitation of results. The other management activity is the one that takes place within the project, and which must ensure that the project objectives are attained, and that participants from different activity areas have a common set of objectives.

The BS is also used by the management when authorising projects, so that they have an overall view of the resources which have been set aside and allocated priorities for the purpose of carrying out the project.

The contents of the BS can be a more or less complete and certain definition of the following:

- Targets:
  – the management's objectives for the project and the overall strategy which the project is a part of

138

- the commercial targets which are to be attained
- sub-targets or strategy interpretations for the activities of marketing, sales, design, production, purchasing, quality, finance and any others which could be of significance for the project
- product-related objectives (specified in the Product Specification, see Section 6.7)

- Expectations:
  - what do we – at present – believe the project will lead to?
  - what sales arguments will we use when the product is a reality?

- Prerequisites:
  - the conditions which must be satisfied for the project to start, i.e. definition of the project's management, organisation, interest group, reference group, suppliers, priority and affiliation
  - significant decisions, if any, on the way in which the project is to progress, its strategy, checkpoints, parallel activities, handling of risks, confidentiality, customer contacts, and so on

- Means:
  - the resources which, on the basis of budgets and plans, are authorised to be put at the project's disposition
  - the organisation and the personnel who are authorised to take part in the project
  - the departments, workshops and laboratories which are authorised to offer services to the project
  - the managerial and organisational conditions which have been agreed on (departure from procedures, norms, decision channels, and so on)

Note that a number of these provisions are really of no significance for the course of the project or its result. The central idea in the BS is the interplay between the project specifications given above and the financial specifications, which may contain the following:

- The commercial specification:
  - potentials: volume of sales, price, profit versus risk, and investments
  - market: price conditions, competitors, trends, relation to one's own products, marketing strategy, sales features
  - product: product strategy, 'the ideal product', rough specification, the product being replaced, cost targets and structures, determining principles (if any), components, relationship to other products, variations
  - production: production strategy, targets, form, type, volume, development in volume
  - project targets: the most important factors in the above, results, deadlines, conditions

- The financial specification: (key numbers)
  - income: sales price, numbers sold in each period
  - production costs: product value, added value, investment in machinery per component, degree of in-house production

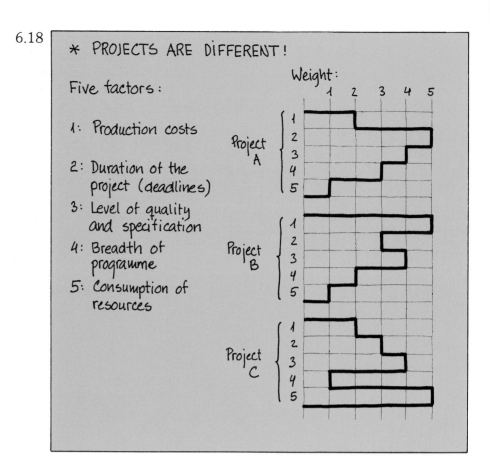

6.18

* PROJECTS ARE DIFFERENT!

Five factors:

1: Production costs

2: Duration of the project (deadlines)

3: Level of quality and specification

4: Breadth of programme

5: Consumption of resources

- sales costs: amount of information material per product, degree of service per customer
- finance: maximum requirement for capital, degree of in-house capital coverage
- rentability: minimum returns, degree of covering, yield
- liquidity: maximum investment per period, earliest time at which expenditure will commence

Working out the BS can appear to be a formal matter, and there is a risk that it turns into a mere formality without real significance. But it is important here to work from the basic belief that all projects are different and that therefore it is important to use the BS to define just exactly those targets and pieces of information which underline the special features of the project.

Valbjoern [31] has suggested a method for fixing the priorities associated with important project conditions. The following aspects of a project are normally important:

- Production costs.
- The requirements for keeping to deadlines.
- The quality and specification level.
- The breadth of the program.
- The consumption of resources.

If the project goes wrong, it will be necessary to depart from the agreed values for one or more of these five parameters. Before the project starts, one can decide, for example, by giving them priorities from 1 to 5, which order of priority these parameters are to have in the current project (see Fig. 6.18).

It is a feature of Valbjoern's five points, and also of the BS, that the means are more important than the final result. It is important to make decisions about important factors related to the project's objectives, assumptions, life and death, before the project starts, so the good projects get wind in their sails and the poor ones are killed off before too many resources are used up. This is done on the basis of assumptions which the project group are also aware of.

6.7 The Product Specification (PS)

The development of a product has three ingredients:

- A marketing activity, i.e. a system of people, financial movements, sales brochures, packaging, sales agents, business premises, exhibitions, means of transport, service workshops and so on, which provides for the sale of streams of products.
- A production, i.e. a system of people, financial movements, purchasing activities, materials, subcontractors, pre-processing and assembly equipment, equipment and routines for control, internal transport systems and so on, which provides for the creation of a stream of products.
- A product, which must have properties and features which make it suitable for production and sale, which make the costs associated with these activities low, and which correspond to the needs of the customer.

Ideally, three results must thus be created during a product development activity, and a specification must be formulated for each of the three desired results. The Business Specification (BS), which we have just discussed, must unify these three results into one business operation. However, it is traditionally the case that existing sales channels and production equipment are used for new products, so that at most marginal changes are made in these areas.

It is therefore normally only necessary to produce a basic specification for the product, although one must be aware of the possible usefulness of specifications for marketing and production, if the aim is to create new results.

A Product Specification (PS) is a specification of that product which we believe to be the best basis for the current business operation. But this specification cannot be formulated in the same way as an existing product is specified, i.e. as a description of a long series of factors which are advantageous and useful, and which are demanded by the user; this in many companies is called a design specification. In development work, what we need is an open formulation of the problem, which gives us freedom to exploit the design degrees of freedom (see Section 4.3). And we need a way of expressing what it is about the ideal product that we like.

At this point, many big mistakes are made in industry. At a meeting of the product committee, you make up a product specification with innumerable requirements, and afterwards wonder why the result of the development project is identical to the company's existing products. Could you expect anything else?

• BASIC SPECIFICATION FOR COIN SORTER

BASIC SPECIFICATION		PROJECT:	EDITION:
FACTORS	DEMANDS/SPECS.	PROPERTIES	REMARKS
Design	Feasibility study: 6 months	Low time consumption / min. physical dims. / few components	
Production	NN-Ltd. technology	low production price / few bought-in components, special materials / simple assembly / designed for ease of assembly	
	Quantity $1-2 \cdot 10^n$ over 5 years		
Usage – input	1-5 types of coin circular / one feeder / diameters: 17-35 mm / thickness: 1-3 mm / average power: 25 mW / current 15-18 mA	simple adaptation to different set of coins	Wish: / No mechanical replacement / vertical slot on front of coin telephone, perhaps on top. What about 5-sided coins? / Thickness may be down to 0.7 mm
– output	two exits: accepted/ rejected / electronic signal for coin's value		
– function	test on immobile coin: / – diameter / – thickness / – conductivity / – permeability / sorting accuracy > 95%	as large as possible / dependable	
	refuse/divert false coins	reliable	
– occasional operations	sorting speed > 1 coin/sec	rapid sorting / easy servicing / easy to clean and remove foreign bodies	how often? modules? preset?
– environment	as for NN-Ltd.'s outdoor coin telephones	vandal-proof	

Shown above is a basic specification for a feasibility study on a coin sorter. The task is formulated in a relatively open manner, as the desire is to see new possibilities and principles. If the project gives results, a new basic spec. is worked out for the implementation project, with tighter constraints and a set of regulatory and standardisation requirements. (From an M.Sc. project carried out in collaboration with GN Telematic Ltd [44].)

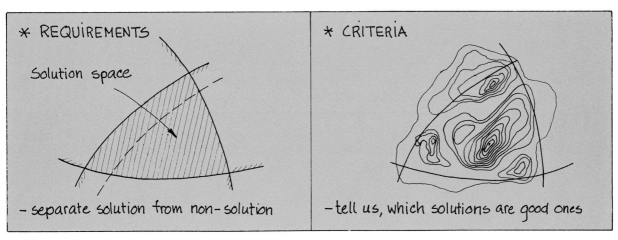

* REQUIREMENTS	* CRITERIA
Solution space	
– separate solution from non-solution	– tell us, which solutions are good ones

6.19

Figure 6.19 illustrates how requirements are used to put limits on the solution space, and how criteria are used to tell us where the good solution is to be found. We know all about this sort of thing from when we employ engineers: "Higher National Certificate in Process Engineering, three years' experience, knowledge of major foreign languages, needed for our department in Copenhagen." By formulating these requirements, we have selected a certain number of possible solutions, but we need to interview each candidate and to look at his diploma, in order to find out what abilities he has: Good exam results? Useful experience for the company? French to grade A, but German to grade E?

The things which are formulated as requirements and criteria in the product specification must of course be factors which have considerable influence on the correct solution, and which make the design process converge. It is no problem to fill the specification with regulatory requirements and speculative details, but we can safely keep them for later phases.

The content and degree of detail in the product specification must be adjusted to suit the number of design degrees of freedom which one wishes to exploit, that is to say, how radically new the solutions are to be (see Fig. 6.20). But which properties should be specified? The main principle is:

The product must possess properties which make it the best possible product at all stages of its lifetime

From the life cycle of the product – design, production, sale, use and destruction – we can therefore identify a number of factors which affect the form of the product. Each factor, such as lifetime, can thus lead to a requirement, Fig. 6.21 ('at least 12 years') and some criteria ('as long a lifetime as possible').

At the start of the project, a product specification should be worked out on the basis of these principles. The document should contain:

- Requirements/specifications for the product, i.e. the fixed and unavoidable requirements which must be applied to the solution and to the specifications which are written down in

143

6.20

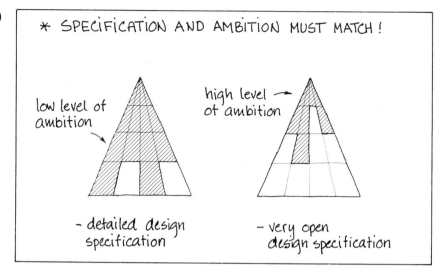

* SPECIFICATION AND AMBITION MUST MATCH !

low level of ambition

high level → of ambition

- detailed design specification

- very open design specification

advance for the product. Requirements and specifications must clearly separate solutions from non-solutions when a choice has to be made.

- Criteria, i.e. those properties or qualities which one tries to attain in the product, and which demonstrate that it is a good product. These criteria are to be used to separate the good solutions from the poor ones.
- Desirable features of the product, i.e. features, details or properties of the product which positively contribute to the value of the product, but which you don't directly want to expend any effort on.
- Remarks in the form of open questions or comments, which turn into requirements or criteria as the design process proceeds and greater insight into the problem is attained.

The product specification involves the management and the project leader, who must agree about its content as a description of the result of the project, the project group, who must see to it as a group that the product is in accordance with the PS, and finally the individual project participant who uses the PS as a target. Thus the PS is used both by the management, for directing the project (result management), and by the project leaders, for directing the efforts of the participants.

The content of the PS will, as mentioned, depend considerably on the level of ambition of the project. The PS is worked out gradually during the development phase of the project, and the aim is to finish it off at the initiation of the elaboration phase (see Fig. 6.22). Note that the PS is not, like the BS, 'complete' from the start, but that it gradually gets completed in step with the accumulation of knowledge. The BS becomes gradually more specific in step with the accumulation of knowledge.

The content of the PS can be as follows:

- The problem specification, which comprises the objectives of the task, its limits, activities, and so on.
- The functional specification, i.e. the sub-functions of the product and the conditions for these sub-functions.
- The construction specification, i.e. the relationship of the product to the system which it is to be a part of, and the

144

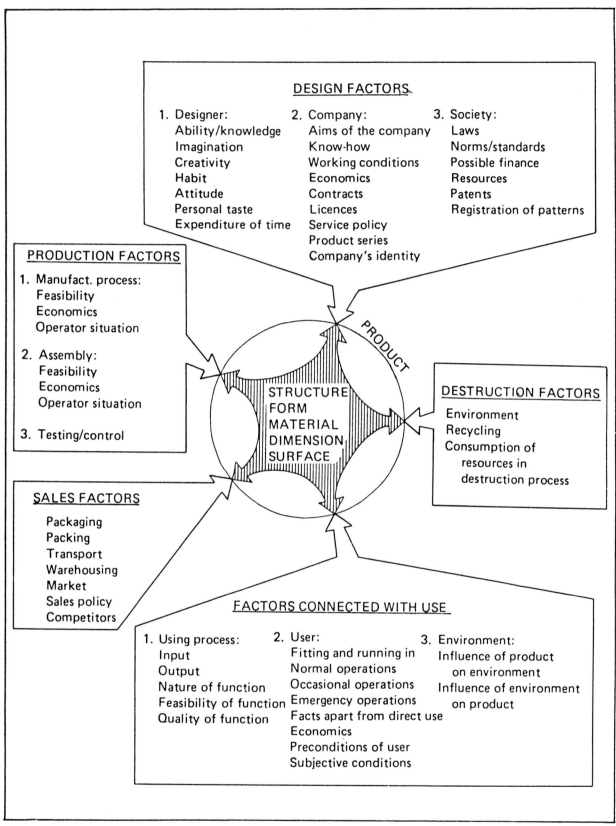

DESIGN FACTORS

1. Designer:
   Ability/knowledge
   Imagination
   Creativity
   Habit
   Attitude
   Personal taste
   Expenditure of time

2. Company:
   Aims of the company
   Know-how
   Working conditions
   Economics
   Contracts
   Licences
   Service policy
   Product series
   Company's identity

3. Society:
   Laws
   Norms/standards
   Possible finance
   Resources
   Patents
   Registration of patterns

PRODUCTION FACTORS

1. Manufact. process:
   Feasibility
   Economics
   Operator situation

2. Assembly:
   Feasibility
   Economics
   Operator situation

3. Testing/control

PRODUCT

STRUCTURE
FORM
MATERIAL
DIMENSION
SURFACE

DESTRUCTION FACTORS

Environment
Recycling
Consumption of
   resources in
   destruction process

SALES FACTORS

Packaging
Packing
Transport
Warehousing
Market
Sales policy
Competitors

FACTORS CONNECTED WITH USE

1. Using process:
   Input
   Output
   Nature of function
   Feasibility of function
   Quality of function

2. User:
   Fitting and running in
   Normal operations
   Occasional operations
   Emergency operations
   Facts apart from direct use
   Economics
   Preconditions of user
   Subjective conditions

3. Environment:
   Influence of product
      on environment
   Influence of environment
      on product

6.21 (after(5))

division of the product into known subsystems, if any. This includes the use of agreed components or principles, if any.

- The situation in which the product will be used, i.e. a description of the use of the product, input/output, operation, man/machine relations.
- The quality specification, i.e. a definition of external properties, and the company's level of quality requirements, testing, product responsibility, approbation requirements, and so on.

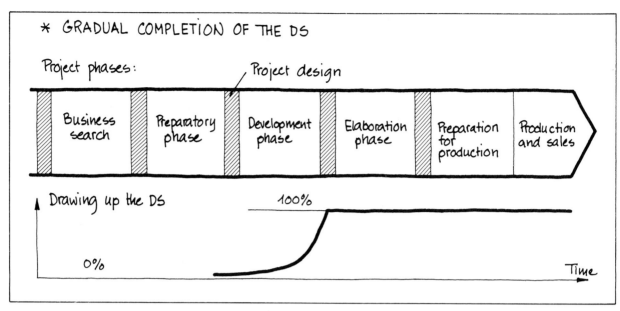

- The sales specification, i.e. definition of the sales philosophy behind the product, important sales arguments and features, sales-related requirements as to packaging, making the product ready for delivery, information material, distribution, service, and so on. In addition, there may be a question of variations, relationships to existing products, and perhaps important competitors and the product's position in relation to them.
- The production specification, i.e. the total volume of production, the future annual production numbers, purchasing questions, and perhaps already defined production conditions (processes, assembly, quality control).

The PS is a target description, and should therefore not contain information about realisation of the target, unless it is a part of the exercise that known principles, components, materials or designs should be used. The PS should be formulated in an operational manner, as one must presuppose that evaluation of, or disagreements about, the results obtained are to use the PS as their starting point.

6.8 Planning the project

Planning in a development project has two main purposes: creating a tool for communication and coordination among those who have contact with the project, and acting as a preliminary demonstration of the project's (and thus of the business operation's) feasibility.

The project leader should be the driving force behind the planning of the project...

...because there is considerable effect to be had from the overall view and the clarification of problems, which comes out of working with the planning of the project, and because the project plan, together with the project's BS (Section 6.6) makes up his area of contact to the company management.

We shall focus here on a planning method which takes the special nature of product development into account, and which satisfies the company's requirements for the inclusion of relevant strategies. This form of planning is called ABC planning, because definite plans are formulated on three levels.

The A-plan (strategic, long-term)

The A-plan reflects the events or states which are of significance for the project's relations to the company or the world around it. Depending on the nature of the task in hand, the plan may contain:

- The times set for keypoints (Section 6.4), or any specific checkpoints (e.g. investment in trial equipment, initiation of marketing trials, start of pre-production runs).
- The deadlines fixed for operational sub-tasks (e.g. demonstration of function, definition of regulatory clauses, choice of trial customer).
- The approximate involvement (in time and quantity) of marketing/sales, design and production activities.
- The approximate involvement of the company's special facilities (e.g. testing labs, model workshops, experimental facilities).
- The times set for making contact to customers, taking part in exhibitions and so on.
- The relationship (in time and content) between the project and any following projects.
- The project's relationship to technology development projects.
- Any strategically determined parallel development activities at project level.
- The project's relationship to other product development projects with marketing or technological areas of contact.

The A-plan (Fig. 6.23) is formally created by transformation of the Integrated Product Development model (Section 6.3), in a way which depends on the nature of the task and the company's resources and objectives, into a plan which expresses the overall project strategy:

> The A-plan shows the project's relationship to its surroundings

But it also reflects the company's attitude to the balance between risk and ambition, to internal interactions, to the exploitation of resources, to time scales, and so on.

The A-plan contains elements which are repeated in the project's BS. Most of these depend, just as in the case of the BS, on knowledge which is normally available at management level within the company. Planning will to a great extent take the form of a keypoint plan [32], but a diagrammatic or verbal presentation form

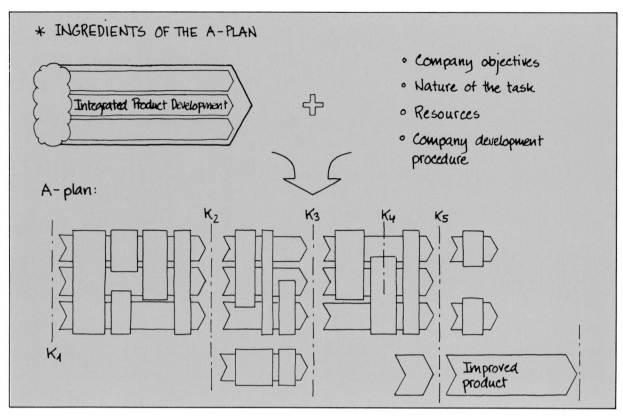

The figure contains:

* INGREDIENTS OF THE A-PLAN

Integrated Product Development

+

○ Company objectives
○ Nature of the task
○ Resources
○ Company development procedure

A-plan:

K_1 K_2 K_3 K_4 K_5

Improved product

6.23

can be useful, where it is a question of showing interactions or relationships. The A-plan normally, even at the start of the project, covers the entire lifetime of the project, but more details are added as time goes by. Like the other plans, it will change as a consequence of changes in the world outside, or as a result of failure to achieve particular results during the project.

The B-plan (tactical, medium-term)

The B-plan reflects the most important tasks and partial results which are to be found within the framework of the project (including what sub-contractors have to deal with). It shows the relationships between these tasks, even across departmental boundaries in the company.

The B-plan shows the project's integration

Depending on the complexity and size of the project, the plan may also contain:

* Resource frameworks.
* The allocation of responsibility to persons, groups or departments.
* The use of methods.
* The project organisation.
* Special reporting or project control requirements.

148

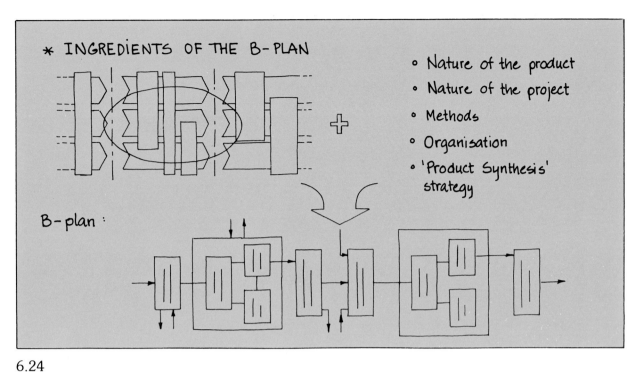

In order to plan at the B-level, the project leader must know:

- The A-plan.
- The level and degree of novelty of the product technology.
- The level and degree of novelty of the production technology.
- The level and degree of novelty in the market and in the marketing technique.
- How the product is divided up into subsystems.
- The type of production.
- The principal areas of interaction.
- The desired degree of safety or risk in the project.
- The policy with respect to subcontractors/in-house production.
- The company's special requirements for clarification on specific points.
- The essential things to be demonstrated in the project.
- Problems which can upset the project.

The B-plan (Fig. 6.24) is created from the A-plan by considering questions of tactics and methods, based on knowledge of the nature of the product, and using Product Synthesis (Section 6.2) as the principal structuring technique. Other important keys to defining the structure are (in different phases) the division of the product and its production into subsystems.

It follows from the above that the B-plan is normally not complete at the start of the project, as certain factors of significance for the plan are not clear until the project has been going for a while. The plan should, however, be available in detail for the current phase (up to the next keypoint or evaluation point), and should be available in outline form for the following phases.

The details incorporated into the B-plan must consolidate the planning of interactions and results which appears in the A-plan, or in the framework for the project's resources and their allocation.

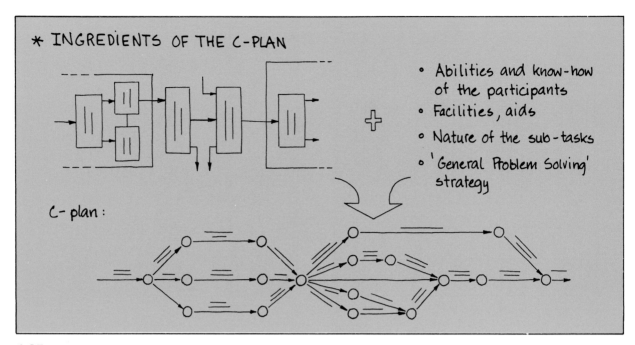

* INGREDIENTS OF THE C-PLAN

- Abilities and know-how of the participants
- Facilities, aids
- Nature of the sub-tasks
- 'General Problem Solving' strategy

C-plan:

6.25

The C-plan (operational, short-term)

The C-plan reflects the individual tasks which are specific to particular persons or pieces of equipment, the resources and timescales associated with these tasks, and any mutual dependency which they might have:

The C-plan is the project participants' map

In order to plan at the C-level, the project leader must know:

- The B-plan.
- The available personnel resources.
- The competence of each individual participant.
- The breadth of view and flexibility of each participant.
- The available equipment and its capacity.
- The methods to be used.
- Those sequences of activities which are subject to time constraints.
- The prerequisites related to each activity.
- Areas of uncertainty for the creation of results.

The C-plan (Fig. 6.25) conforms to the structure of the B-plan, with the details filled in by the use of General Problem Solving (Section 6.2). The C-plan is normally, but not necessarily, complete for the current phase of the project. The plan for this phase can well be affected by matters which are cleared up in the phase itself.

The details incorporated into the C-plan must consolidate the B-plan. If this leads to conflicts, then adjustments must be made to the B-plan, and possibly also to the A-plan or to the framework for the project's resources and their allocation.

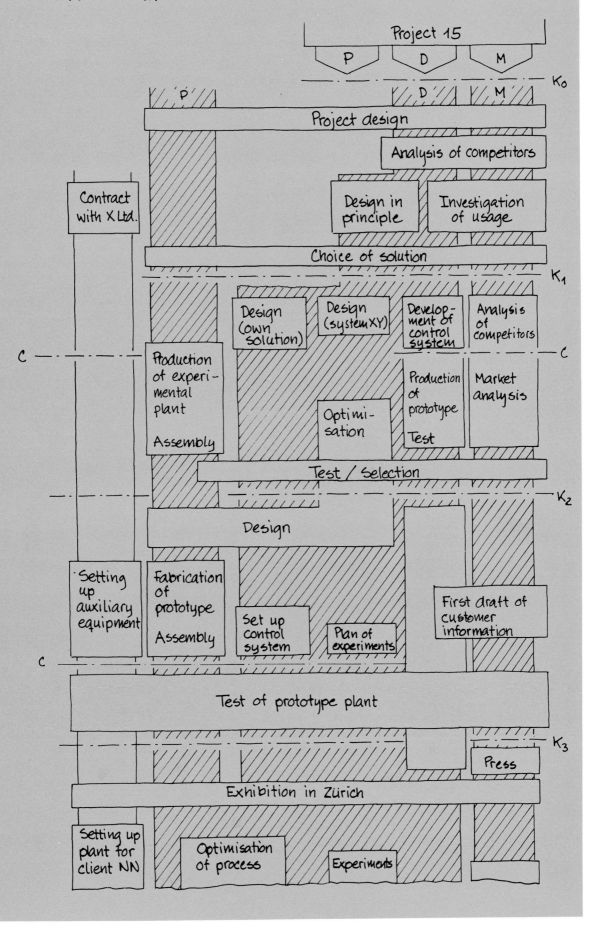

* A-PLAN FOR PROJECT 16

151

The form of the plans

ABC planning takes place in a top-down manner, first the A-, then the B- and finally the C-plan. However, this sequence is, as previously noted, iterative, in the way that the incorporation of details into the B- and C-plans can reveal unrealistic requirements for results or deadlines, which must lead to adjustments in the plans at higher levels.

While the contents of the A-plan are in the nature of wishes or demands which the project group has to fulfil, the contents of the B- and C-plans can be manipulated by the project leader. Putting the necessary activities, resources, results, and so on, onto these plans is a process which is complicated and difficult to programme, and which should also be iterative. Switching between forward and backward planning is a useful tool for setting up temporal relationships. A good technique for ensuring completeness is to note down all relevant activities and associate the corresponding partial results (or 'states') with them. Then you should note down every imaginable partial result/state, and associate all the activities which can lead to this result with it. In this way you have a sort of double-entry bookkeeping which can be checked, and a complete picture can be obtained.

The concrete graphical representation (network, cyclogram, etc.) chosen for each plan must depend on the task in hand, and in the opinion of the authors there is no technique which is especially to be preferred. We refer you to the literature on project planning and control (e.g. [33]) for a presentation of known and tried techniques. Attention should be drawn, however, to the fact that, even if there is a certain similarity between the structure of ABC planning and traditional planning levels – main plan, phase plan, task plan – you will discover that:

> The literature about project planning and control deals only to a limited extent with product development

Most of the literature deals with projects which are mere planning projects, and these differ from most product development projects in that:

- From the very start of the project it is possible to have detailed knowledge of the individual project activities and their relationships to one another.
- The uncertainty associated with the individual activities is a question of their consumption of resources or time, and not of their feasibility or of the quality of the result.

Most of the planning tools which are offered have been created to suit this project profile.

Is product development like playing pin-ball? With sequences which occur too rapidly to be planned properly, with opportunities which are lost because your reactions weren't quick enough, and with things which suddenly and inexplicably succeed better than expected, so that the counter whizzes up? At any rate you don't win every time, and you get no clear warning of whether you are in for a big payout – or whether all your stakes are lost.

7

Assessing chances...

SUMMARY

In this chapter three factors which are crucial for the success of the planned business operation are discussed. First, the internal risk, i.e. the risk of the project not being successful in creating the product, production system, sales system, service and so on, which the specification requires. Second, the external risk, i.e. the risk of the business operation which has been specified not being able to survive, because the need is not there, the production system doesn't work, or thirdly, the competition which forms the background against which the product, its production system, marketing, service, and so on, must clearly stand out.

Finally, we look at the handling of risks: the adjustment of levels of ambition and willingness to take risks to suit the task in hand, and the reduction of risks as much as possible by strategic, tactical and operational initiatives.

7.1 Internal risk

In development activities there is always a certain risk that the target is missed, and the project to a greater or lesser extent turns into a failure. To fail to reach the target has consequences with respect to the company's financial position, goodwill, market position, and so on. In order to have a concept of risk which we can handle, we shall in what follows consider risk as being the probability of failure times the consequences:

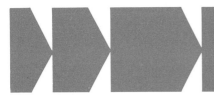

The internal risk is the probability that the project group does not reach the target, multiplied by the consequences of this

The probability that the project will not enable us to create a product, production system, sales system, and so on, which correspond to the desired specifications depends on the ratio between the effort put into the project and the level of ambition[34]. The effort involved can of course be described in economic terms, but many non-quantitative factors also contribute:

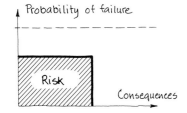

- How is the project staffed? Do the participants have technical ability and competence in taking part in projects? Does the group have the necessary breadth of competence? Does the leader have the necessary experience? (Fig. 7.1)

- What priority does the project have? Are the participants free from other calls on their time? Has a suitable framework been set up for the project group? Do the participants have backing from their respective departments?

- How is the project carried out? Are resources used as well as possible? Are things done flexibly, with opportunities and interactions being exploited? Does the management give the right signals at the keypoints? Are there decision crises? Do things run out of steam?

These factors, and a number of others, affect the effort put into the project and decide the power which can be brought to bear on the task in hand[35]. This becomes apparent when the project group gradually creates the product in the course of the project by generating alternative solutions and making selections. If there is no breadth or quality in the creation of these solutions, the design opportunities may be wasted. If an overall view is not available in the situation where choices have to be made, based on criteria derived from marketing and production considerations, then this can lead to sub-optimal choices which weaken the business operation as a whole.

Those specifications which turn up in the project do not in themselves reflect the ambitions behind the task in hand. Many factors contribute to setting the level of ambition:

- The degree of novelty with respect to product technology, production technology and market

156

7.1

* THE INTERNAL RISK

This is what we are throwing in: — and this is how big the task is:

— can we reach our goal?

- The level of specification in relation to known products.
- The number of variations which are to be dealt with in the project.
- The number of participants, departments, special facilities and subcontractors involved.
- Pressure of time or work.

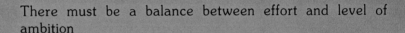

There must be a balance between effort and level of ambition

The consequences of missing the project target, in the sense that the specifications are not met, are that the effort put into the project is lost. Thus it is important to demonstrate the project's feasibility as early as possible, so that one can consider stopping it while the consequences of doing so are modest. The consequences increase in steps, corresponding to transitions between the main phases:

- If a project is stopped after the business search phase, then very little is usually lost.
- If a project is stopped after the development and design phase, then possibly a considerable amount of work goes wasted, but...
- If you have to stop a project after investing in production equipment and perhaps buildings, then the consequences can be fatal.

As most companies have the attitude that they can live with a risk of a certain magnitude, it should be true that:

The probability of failure must be reduced in step with increases in investment in the project

Sometimes one chooses, when problems arise in a project, to change the specifications so that they match what can be attained, rather than keeping to what was originally thought to be necessary for the business. Naturally, this is done in the hope that the business operation will be able to survive with the reduced specifications – a belief which will tested when the product comes on the market.

Once the product has been launched, the consequences of failure are much worse than during the course of the project, so:

> Stop a failed project before the product is put into production and sale!

Obvious, isn't it? But in many companies stopping a project means personal failure, perhaps even loss of jobs, so nobody has the strength of mind to stop a project.

7.2

7.2 External risk

The probability of failure

The probability of the product being a failure when production and sales have started, if we leave the competition (which is to be dealt with in Section 7.3) out of the picture, can be considered as made up from two components:

- The probability that the business concept is wrong: there is no market for the goods.
- The probability that the specifications are unsatisfactory in practice, so that sales fail to materialise.

> The external risk is the probability of the product not reaching the expected sales figures, multiplied by the consequences of this

The business concept is wrong

Important requirements or desires have been overlooked when the specifications were worked out. The need has been perceived in a directly incorrect manner, it alters as the project proceeds, or there are social, cultural or legal barriers which have been overlooked.

Where this is the case, the user's situation has been too unclear for marketing/sales and design, and at the same time the marketing possibilities have been incorrectly proven during the project. Lack of clarity with respect to the user's situation is naturally the most obvious explanation when it is a question of radically new products, where there are no precedents and/or competitors which can be studied, but it is also possible to misinterpret the user's situation in more run-of-the-mill projects (Section 8.2). Thus demonstrations of acceptability using user panels, pilot customers, trial marketing runs, and so on, are important means for reducing the probability of failure.

Failure during marketing can also be due to the fact that the company's intentions with the product do not reach out to the customer. Perhaps incorrect or unclear arguments are used, or the line of thinking is quite foreign to the consumer. This type of failure lies on the boundary between external and internal risk, but basically the fault is the company's.

The specifications cannot be met

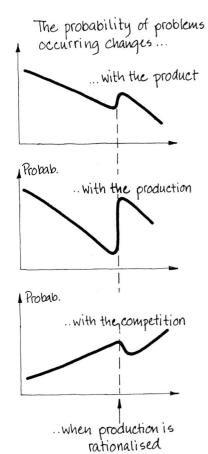

The probability of problems occurring changes...

...with the product

Probab. ..with the production

Probab. ..with the competition

..when production is rationalised

The specifications, understood in the broadest possible sense as requirements and desires with respect to sales, service, production and so on, are not finally met when the product development project terminates. Perhaps the product has to work under different running conditions than those which have been tested in the laboratory or in test rigs, with the result that the product specifications cannot be maintained under working conditions:

- The running conditions for the product are tougher than expected.
- The product is incorrectly handled by the user.
- The product is used for other purposes than those for which it was intended.
- Dimensions or materials are changed within the limits of the specification with the result that the product's function, reliability or lifetime is affected.

The production system must, once established, demonstrate that it lives up to the requirements and wishes which are made of it, and there may be uncertainty with respect to a number of factors:

- Quantity: Can the cycle times and efficiencies for the individual production machines be maintained? Does the interaction between production units work well? Can the subcontractors keep up?

- Reliability: Is the reliability of the individual machines as predicted? Can (and will) subcontractors deliver on time? Can critical tools and machine parts be obtained as fast as presupposed? Does the control of component and material flows work well?

- Quality: Can dimensioning and material specifications be maintained? Do the tools last as long as presupposed? Can the quality of parts obtained from subcontractors be controlled?

- Flexibility: Can we keep control of variants of the product? Do changeovers between variants work as well as expected? Are minimum and maximum loads handled cost-effectively? Can the production system be upgraded to increased volume, a greater degree of automatisation or new materials and processes?

- Finance: Can the product's production price be maintained? If not, can the sales price be maintained?

Corresponding open questions can be posed with respect to the sales and service systems: Does the message get out? Can they deal with the quantities involved? Can they deal with the interactions with previous products? Are they efficient at dealing with a new product? Can variants be handled? Can they deal with the new processes and new technology involved?

Consequences

A significant factor in the consequences of failure after the launching of a product is that it takes place in view of the public, and may have a negative effect on the customer. The spectrum of consequences is, of course, very broad, from slightly reduced profitability to recall of the product with loss of goodwill for the company and its other products as a consequence.

In the period when the product is on the market, the consequences of failure increase in jumps, in connection with investments in expansion or in rationalisation of production.

7.3 The competition

The product's chances of success depend on the competitive situation in which it has to exist. The competitive situation will be influenced by the manufacturers' size and number, and by the features – such as the product's function, price, reliability, lifetime, sales and financing conditions – on the basis of which the competition is taken up (Fig. 7.3). For the management, it will normally be an important part of the information on which their decisions are based, that the position with respect to the competition is known.

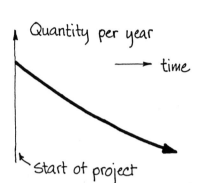

"Why do the marketing department's expectations for sales figures always fall proportionally with the time for which the project has been running?"

(Head of R&D in a Danish industrial company, 1984.)

7.3

The competitive situation in relation to product development is handled in general terms by the management through their selection of a product policy, and through the continuing activities of planning products and giving them priorities. The intention with a product development project is often to obtain a competitive advantage, so:

> **The management's perception of the competitive situation must be known to the product development project**

Here it can, amongst other things, turn up in the Business Specification as an answer to the questions:

- Who are the most important competitors in relation to this product?
- What are to be the most important competitive features for the product?
- What will be the critical things in this competition?
- How large a share of the market must the product obtain for the company?
- What general development tendencies are to be found in the competitive situation?

These pieces of information should be continually checked and updated, since they are crucial for whether the project group can reliably analyse or exploit the competition in the course of the project.

> **The competitive situation is in continual development and must be monitored all the time**

Competitor analysis

The analysis of competitors is an important tool which is not limited to a single phase of the development process, but which can be used to advantage in many situations. The objective of the analysis

7.4

7.5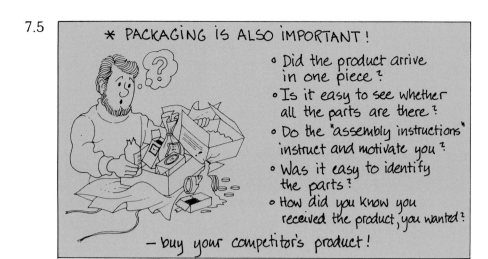

is to perceive facts which can rationalise and strengthen the product development activity, and here it is the case that:

> The dividends from competitor analysis increase with knowledge of the problem

This is true both with respect to the situation in which the product is used, and with respect to product and production technology. Competitor analysis should therefore take place in an interaction with other product development activities, and one cannot put a special management group onto the job.

Competitor analysis can contribute to development work in all those areas which characterise the product, and thus one should look at the product from several points of view (Fig. 7.4):

- The market: In the initial phases, information should be accumulated about the market situation. The competing products reveal how the market is divided up with respect to usage, price level and quality.

- Usage: The situation in which the product will be used is the purchaser's interface to the product. During competitor analysis, the designer should act as user and notice good and bad properties from the user's point of view (Fig. 7.5). Note

7.6

that the competing products may have created a standard (and expectation about the price, function, design or whatever) which it can be difficult to ignore.

- Possible solutions: A competing product forms a concrete example of a way in which the design degrees of freedom can be combined, and can thus contribute to our general idea of the solution with respect to processes, functions, structure, form, materials, and so on (Section 4.3). Analysis of processes and process parameters (including sensitivity analysis) will often be an important element in competitor analysis (Fig. 7.6).

  Drawing up process diagrams (the way in which the process proceeds in time and space) can be a help to the creation of an overall view of the main alternative process possibilities. Functions (and means for realising them) can likewise be systematised, so that one obtains both a general view of the solution space and a key to systematic ordering of the competing products. When analysing the form and choice of materials in the product, one should be aware that these are dictated by sales requirements as well as production requirements. The product can also be influenced by conditions within the company, as for example:

  - that the production system must also produce other products
  - that the product has certain components in common with other products
  - that certain components, processes or materials are particularly cheaply available

- Production (process and assembly): Analysis of process and assembly conditions will be able to give information about the competitors' level of ambition with respect to production volume, variations, level of quality and so on, and will be able to give inspiration to the collaboration between design and production activities.

When carrying out competitor analysis, it is important to remember that the products which one looks at have been developed several years ago, and that – at this very moment – the competitor is busy working on a new model, based on the experience he obtained with the production, marketing and service of the current product:

It is not this product, but its successor that you have to compete with!

Competitor utilisation

The competing products will have a greater or smaller degree of similarity with the product you yourself are developing. This similarity should be exploited:

163

- Test systems and working models: You will often get through the stage of test systems and working models more quickly and cheaply if you use more or less cannibalised products from your competitors.

- Components and subsystems: In prototypes or small preliminary runs, you can include components or subsystems from your competitors, if these are critical with respect to production technology, production time or production finance (Fig. 7.7).

- Substitution: If a subsystem for your product can be bought from your competitor, you can use this during an interim period. In this way you postpone a part of the development load, and thus reduce the risk associated with the project.

- Marketing trials: If you want to try out the reaction of a limited market to your product, you can perform your trial using purchased examples of your competitor's product. You can also market competing products simultaneously with your own, to test differences in the market's reaction.

7.4 The handling of risks

The handling of risks during and subsequent to the project is based on knowledge of factors related to the internal and external risk, and to the competition. This is reflected in a number of situations in which risks are dealt with during the project (regardless of whether one is aware of it or not!). We shall now look at a number of these factors.

Project planning

At the strategic level (the A-level of Section 6.8), risk is minimised by:

- The general allocation of resources and staffing levels.
- The incorporation of parallel sequences of development.
- The use of 'technology packages' and off-the-shelf results.
- The incorporation of pilot customer contacts and trial marketing runs.
- The division of the development task into several rapid, market-oriented product development sequences.

At the tactical level (the B-level), which shows the most important project activities and their interrelations, risk is minimised by:

- The allocation of resources to critical tasks or ones which are in difficulties.
- The use of known results instead of doing your own development.
- Working in parallel on several solutions.
- The incorporation of suitable interplay between departments.

7.7

★ USE YOUR COMPETITOR'S COMPONENTS IN YOUR OWN PRODUCT DEVELOPMENT!

- The incorporation of monitoring of the competition.
- The incorporation of regular checks on the development of the market.
- The incorporation of continual trials of the product against the market.

At the operational level (the C-level), which shows the tasks of the individual project participants, risk is minimised by:

- Putting qualified people onto the most critical jobs.
- The incorporation of competitor analysis and utilisation.
- The incorporation of sensible ways of going about problem solving.
- The incorporation of parallel activities during critical tasks.
- Making sure that resources are available for increasing the effort which can be brought to bear along the critical route through the plan.

Specifications

When the Business Specification is being worked out, risk is handled by the adjustment of the ratio between the level of ambition, with respect to the project's target, and the resources which are to be used in the project. This includes making decisions about which factors are to be considered as freely manipulable, and which are fixed (e.g. the use of existing product elements, an existing production system or an existing sales system).

In the Product Specification, risks are likewise governed by the relationship between the level of specification (strict requirements, loose requirements) and the allocated resources. The specifications have the peculiarity that they don't necessarily give the possibility of there being any solution at all!

Modelling

The handling of risks is closely associated with demonstrations of the workability of the project, i.e. those activities where it is shown that the assumptions made about market acceptability, function, user features, assembly and process features, and so on, are correct ones. The proper use of model technique in the form of working models, mock-ups, prototypes, experiments, test series, pre-production runs, etc., makes it possible to demonstrate these things one at a time, and in an appropriate order for minimising the risk. The total proof which is involved in the production of the product, its launching and subsequent acceptance by the customers can thus be carried out in a step-by-step manner:

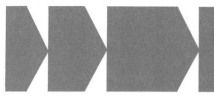

> The planning and execution of modelling activities should be based on a line of thought directed toward minimising risks

... and without narrow-minded, 'reflex' focusing on fixed concepts such as the working model, the prototype, and so on.

Looking for solutions

The search for solutions affects the risk associated with the project because the extent and completeness of the search is a measure of the probability of finding a good solution. Because of this, the search for solutions becomes, perhaps involuntarily, an indicator of the project's real level of ambition.

Evaluation and selection

The way in which you deal with the activities of evaluation and selection, which turn up repeatedly in the course of the project, has a decisive effect on the risk:

- You can choose solutions which are 'certainly realisable', and thus reduce the risk (and perhaps also the potential gain).
- You can make selections on the basis of clearly defined and well-substantiated criteria, or you can guess at a solution.
- You can evaluate and select frequently, so as to attain a cost-effective search through the solution space.

Particular weight must be attached to those evaluation and selection situations which the management (or the product committee) are presented with at the transitions between phases of the project. These are partly a question of approval of the way in

7.8

* THEY'RE ALL EQUALLY DIFFICULT – BUT THE RISK...

which criteria have been utilised and choices have been made, and partly of approval of the work on which these decisions have been based and its reliability. In this situation, the management affects the risk associated with the project decisively when it approves or rejects this work.

We have several times in this book drawn the reader's attention to the way in which risks are dealt with, e.g. in the evaluation of success factors (Section 5.2), in the composition of the project group (Section 5.3), in ABC planning (Section 6.8), in working out the Business Specification (Section 6.7), in choosing a project strategy (Section 6.5) and in performing design reviews.

Willingness to take risks

"But the highest risk products, which account for 30% of all introductions, represent 60% of the most successful new products" [17]

The larger a company is, the less willing it is in general to take risks. This is because there are more people who have no insight into the potential of a project, but are only able to see possible elements of risk within their own areas. The result is wasted opportunities.

The entrepreneurial spirit must prevail, and opportunities must be evaluated both intuitively and systematically – one minute you must follow your intuition, the next you must follow carefully evaluated proposals. The cautious executive or project leader will be exposed by your competitors, whose products will become better. The clever executive or project leader is daring but not rash: he or she takes care to handle risks rather than leaving them to chance.

8

... and exploiting opportunities!

SUMMARY

In project activities, quality is to be created. The properties of the product, its features and positive attributes, must involve an advantage in relation to the competition, and must therefore correspond to the level of ambition of the project, and the effort put into it. The product's positive attributes comprise not only those attributes that the customer values highly, but also those attributes which are reflected in a high efficiency within the company, and which thus lead to low levels of costs.

A prerequisite for the incorporation of positive attributes is that there are some design degrees of freedom available, and that they are exploited. This implies mastery of the techniques of design for economy, design for quality, design for process, design for assembly and design for many other sorts of thing. As examples, we shall look at the process and assembly areas.

Methodical design is the means for handling the design degrees of freedom and therefore plays a considerable role.

8.1 The integrated project

In Chapter 7 we explained how a project's external framework is recognised, created and dealt with. It is important that the project fully exploits and perhaps even breaks out of this framework, so that the best possible result is obtained, and the group feels itself challenged to give its best. A football team which never plays a match deteriorates. If the captain often calls 'back to cover' and never 'forward', then the chances of scoring a goal are small.

Product development is a very complex activity. Within each of the areas of marketing and sales, development and design, and production, many things take place, which themselves are complex, but which also in a complex manner depend on and affect activities and requirements within the other areas.

Three considerations must be taken into account in product development:

- The costs and efficiency of the development project.
- The quality of the project result – i.e. the resulting business and its elements: the market, the product and its production.
- The company's costs, efficiency and development.

The first of these has been dealt with in several chapters, where we have shown how the integration concept can influence the start, organisation, planning, control and management of development projects. We shall consider the second point, the creation of results of high quality, more closely in this chapter, and will for selected areas demonstrate how the integrated method of attack offers great potential for improving the result. The third point, consideration of the company's costs structure, efficiency and development in relation to its development projects, will be dealt with in Chapter 9.

A long series of features characterise the good project result:

- The quality of the product as experienced by the user.
- The price, running costs, maintenance, service, and so on, for the product, i.e. factors related to its life-cycle.
- The situation of the product with respect to safety, product liability, regulatory directives and standards.
- Attributes of marketing, such as efficiency and costs.
- Attributes of production, such as process and assembly costs.
- Attributes of purchasing, such as guarantee of delivery, price and quality.

Thus the incorporation of quality involves two questions of quality: creation of those positive attributes which the purchaser is willing to pay for, and creation of those positive features in the project result which lead to low costs within the company.

In the following sections we shall first look at the concept of quality in general and at those tasks and problems which are associated with the handling of quality during a project and during production and sales. Then we shall consider the creation of quality, which takes place by exploitation of the design degrees of freedom, and which involves a whole series of individual areas of design: design for economy, design for quality, design for process, design for assembly, design for control, design for marketing, design for sales, design for usage, design for maintenance, design for operation and design for disposal. The list is long and perhaps exaggerated. Most of the areas mentioned have not been offered much attention, and we shall refrain from looking at the market-related ones, because they lie outside our area of competence.

We have chosen to illustrate the creation of quality by considering two production-oriented areas, namely the assembly area and the process area.

8.2 Quality

The quality of a product is the customer's perception of and evaluation of the properties of the product, i.e. the properties he or she observes and attributes to the product, together with the set of his or her value norms used as a basis for evaluation of the product. A product sells on the basis of its good qualities, not because it can do anything particular. Indeed, for almost every product there is a long series of completely corresponding or substitutable competitor products, which means that when the customer chooses, the product's function is assumed to be in order, and its qualities come into the picture.

When we talk about the product as a 'moving target' (see page 86), it is in particular people's perception of quality which is liable to change and difficult to make tangible. Thus it is very important to direct product development towards concrete quality targets as given in the Product Specification (Chapter 6) and to ensure that this quality is created and maintained during production, marketing and use of the product.

The quality targets and quality control which must therefore be present within the company are not just to be related to the product, but also to the instruction manuals, data sheets, installation guides, service instructions, spare part catalogues and guarantee arrangements – to name just the elements associated with documentation – and also training, installation assistance or actual installation, maintenance, service, and so on[36].

> It is the product as a whole which must have the right quality

During development, the question of quality is associated with some ideal for the product, but as soon as we attempt to realise this ideal in the form of a working model, prototype, pre-production run or actual production, then it becomes a statistical matter. Components, assembly, setting-up and testing will vary from product to product, which means that the quality deviates to a greater or lesser extent from the ideal, inherent quality. Quality-related activities are therefore of two types: firstly, defining and

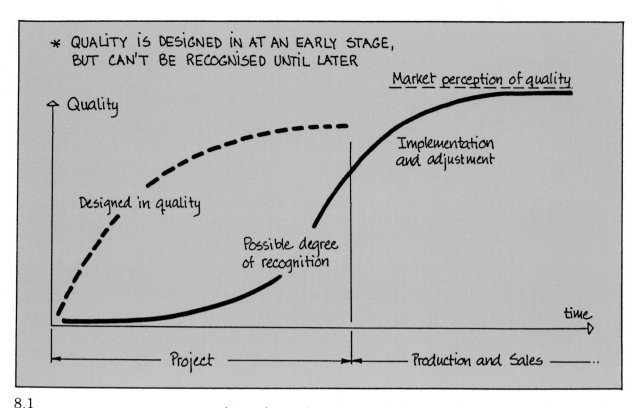

8.1

making the right choice of the ideal quality, and secondly, controlling the achieved quality so it has an acceptably small statistical deviation from the properties of the ideal.

The market's perception of quality is subject to trends such as the following[37]:

- The consumer is becoming more and more quality- and price-conscious.
- Competition is on the increase.
- Reliability and lifetime are continually being improved.
- Immediate failure (i.e. where the product doesn't work when the customer begins to use it) is considered unacceptable.
- Guarantee arrangements are becoming more comprehensive.

Everything indicates that quality is 'designed in' in the same parts of the project as those where costs are determined, but our possibilities for recognising, measuring, checking and correcting faulty quality follow a much flatter and later rising curve (see Fig. 8.1). There are a long series of reasons for deviations in quality:

- There are no design methods which directly lead to the incorporation of quality. We are reduced to backward-step methods, where we attempt to create solutions and afterwards to check whether the quality obtained was the right one. Of course, experience and knowledge of the principles of quality-oriented design play an important role here.

- Those who define the quality and design it into the product don't see it realised and therefore achieve poorer insight into questions of quality. For both the marketing and the development departments, the consequences of quality requirements and the specification of the product cannot be seen in the departments' own activities, but they turn up in

8.2

the preparation for production, in production and in quality control – and, of course, at the customer's! Know-how and experience with respect to quality cannot therefore be built up on its own, but requires integration 'lengthwise' through the project.

- The realistic picture of the customer's needs and quality requirements is looked after or interpreted by a whole chain of people, who speak on the customer's behalf, so that the picture gets distorted. It is important that the development group itself gets the opportunity directly to interpret the wishes of the customer (Fig. 8.2).

- There are conflicts of interest amongst those who are involved in product development. Partly because people in the earlier phases have to solve the later phases' quality problems: this is troublesome and they know too little. And partly because attention to quality leads to sub-optimal solutions with respect to other properties – as exemplified by the contrast between the design department's local view of quality ('the technically perfect product') and the customer's ('a good investment').

- Holistic solutions are 'nobody's' job. As stated previously, quality is associated not only with the product but with all the efforts which go to making a good piece of business, i.e. a satisfied customer and a satisfied supplier. The business has market, product and production aspects, which are often looked after by different people. Thus the overall view tends to disappear, and people 'forget' to tell the development department what features they want to stress in a sales campaign, or 'forget' to incorporate quality control into service or installation, or 'forget' to gear the dynamics of the production system to sales, so that the products can be delivered in time.

- Quality is difficult to control from a statistical point of view. Fabrication and assembly processes create variations dependent on the type of process and equipment in use. New processes are normally a problem, since their variation or capability has to be directed into an acceptable level. So, when the designer chooses a particular design and thus particular processes, he must not just think about costs but also about process stability.

How is designing quality into the product to be handled? Who is to perform this design? In principle, quality-related tasks must be looked after where quality is decided, i.e. where the solutions to the marketing, product and production tasks are created and selected. But unfortunately we are only rarely in a position to recognise and measure quality when it is decided (Fig. 8.1). Thus we have a need for control and adjustment activities at a later stage – i.e. a 'quality task'. The aim of a quality task is to create products and production systems where the costs associated with maintaining quality are minimal. The basic idea here is that quality-related costs arise because we are not good enough at finding design solutions and processes which give the desired result without such costs.

The main problems of quality control are:

- To achieve the quality that the consumer prefers.
- To avoid faults appearing in the sold products, or complaints arising about them, by the creation of more thoroughly tested products which are better adapted to the user.
- To avoid costs associated with control, correction of faults, selection, scrapping, testing and so on, by attacking these problems early and efficiently.

The task of checking and adjustment, which must return the quality to the right level when deviations are observed, is often an explosive issue, since those who perform the checks are seldom the same as those who perform the adjustments. This specialisation gives integration problems and conflicts. The problem is difficult to solve because there is often a considerable time lag between the origination and the detection of the quality deviation. Perhaps the designer has been put onto a new project some while ago, and no longer has any motivation for working on the old one. There are good reasons for keeping project participants on a project for a

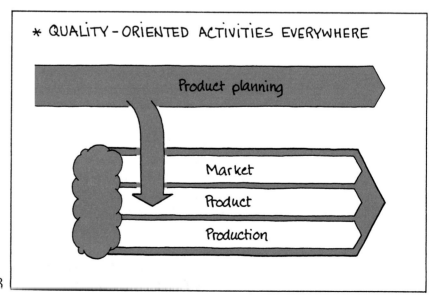

* QUALITY - ORIENTED ACTIVITIES EVERYWHERE

Product planning

Market

Product

Production

8.3

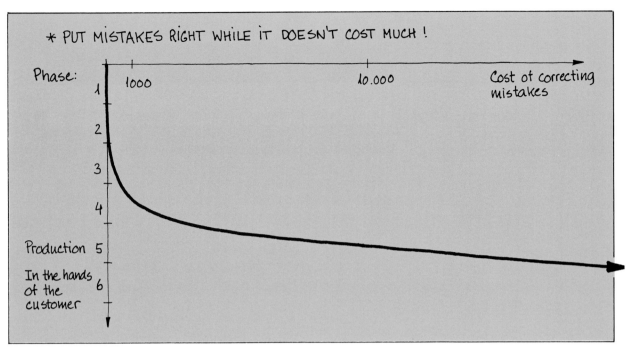

* PUT MISTAKES RIGHT WHILE IT DOESN'T COST MUCH !

Phase:

1

2

3

4

Production 5

In the hands
of the   6
customer

1000   10.000   Cost of correcting
                              mistakes

8.4

considerable time, so they participate in the solution of the problems which are caused, and can accumulate a more long-term insight into the consequences of what they did early on in the project.

The quality problem should be handled in an overall quality control activity [36], and should include all phases and all phase results in an integrated product development and its planning (see Fig. 8.3). Quality-related costs in a quality organisation are normally divided into costs associated with the tasks of prevention and monitoring, and those associated with losses, which include internal and external loss-related costs.

The prevention-related costs are only the quality organisation's costs for control (e.g. for design reviews), while the costs incurred for preventive quality control in marketing, design and production preparation are not included. The influence of these departments on quality is much greater than the influence which can be created through the subsequent quality-related activities. And paradoxically enough, the quality-related costs in these areas should be zero, since:

It doesn't cost anything to do things right the first time

A clear view of what the right way should have been is often first obtained with the help of the brilliant illumination of hindsight, but the costs associated with correcting a mistake rise with roughly a factor 10 every time we move a phase on in the course of the project (Fig. 8.4).

Some important methods which integrate the tasks performed within the three activity areas, and which lead to the exploitation of possibilities, should briefly be mentioned here.

175

Design review is a method which is primarily used in connection with quality control. It is normally performed through meetings held in connection with project keypoints. A panel consisting of experienced designers and representatives from the marketing, production, quality checking, and service departments, possibly with the inclusion of some customers, go through the proposed product (in the form of sketches, drawings, working model, prototype, pre-production run, initial production) and evaluate it from a variety of points of view, such as safety, risk, function, fabrication, quality, maintenance, usage, operation, personal safety, product liability and destruction. The method gives good results, but presupposes that the designers are prepared to accept criticism from outside, and are ready to change the product.

Design for quality is an undeveloped area. As mentioned, there is a very complex relationship between those quantities which are set by the designer, and their consequences in the form of production and quality-related problems and costs which are incorporated into the product. It is important that the designer is acquainted with the principles of design for quality, and knows the different types of quality-related costs which are incurred in the different phases of production.

In Section 5.2 we discussed experience curves for product development. The quality area is a part of this and should likewise be considered as a field in which an experience effect should be built up. An important contribution is to learn from one's own mistakes[38], i.e. to perform retrospective analyses of completed projects with a view to making clear (Fig. 5.5):

- In which phase did we discover a problem?
- In which phase should we have discovered the problem?
- In which phase did we create the problem?

The method presupposes that product development is clearly divided up into phases and tasks, and that the analysis is carried out via a careful diagnosis of causes. Even if a problem can be traced back to the designer's choice of a particular principle or to the product's structure, the designer would not be able to recognise

8.5

176

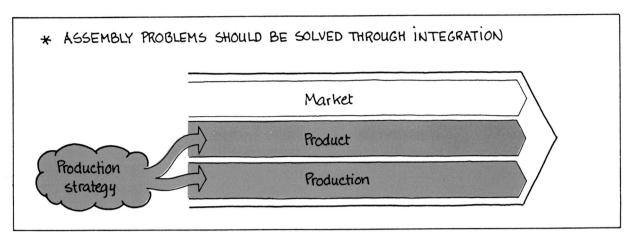

* ASSEMBLY PROBLEMS SHOULD BE SOLVED THROUGH INTEGRATION

8.6

the problem until much later during the design process, when the details of the product had been settled and the quality could be determined. So the diagnosis must be carried out in a realistic manner.

Modelling, in the form of mathematical models, working models, prototypes and pre-production runs, and realistic production results are very important for dealing with quality problems, and therefore require careful planning – both of their execution and of who is to evaluate the results.

8.3 Assembly quality

The rationalisation of assembly is an area which cannot be considered in isolation as a matter of production technique, but must be seen in a wider perspective as a development task which must have clear strategic objectives and be performed in a collaboration between the design and production departments (see Fig. 8.6).

8.7

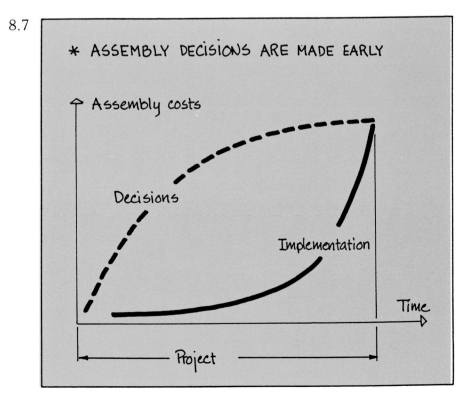

* ASSEMBLY DECISIONS ARE MADE EARLY

Assembly costs are decided at an early stage and come into effect much later, as shown in Fig. 8.7. The designer decides which types of assembly are to be used, and which fabrication processes (with related assembly problems) are to be used for the components. Whether the designer is aware of it or not, a major part of assembly costs is decided when the product is determined. If too little attention is in many cases paid to assembly, and the results of assembly are therefore inferior, then this is often due to:

- Pressure of time and perhaps poor planning.
- Lack of awareness of the importance of assembly as a contribution to costs (40-60% of the time consumption during production).
- Lack of knowledge of design for assembly.
- Organisational problems which inhibit a fruitful collaboration between people from different departments.

The aim of assembly rationalisation must be improvement of the efficiency of assembly, the product's quality, and, since it is a question of interactions between products, operations and machines, improvement of the working environment. These quantities together make up the assembly quality of the product. In order for the designer to be able to handle this rationalisation, he must thus be able to foresee the consequences of his decisions, that is to say:

- Know which design parameters determine assembly quality.
- Be able by design to form the product so it attains a high quality.

This task is very difficult. The relationship between the form of the product and its consequences for assembly is very complex, and there are no methods which can be used directly to incorporate suitability for assembly into the product. We can only go in the opposite direction: create solutions, check assembly properties and choose the best alternative.

The form of components, choice of process and assembly are closely interrelated. Some processes and materials permit the production of composite shapes and can thus be used to give integrated components without assembly, while others permit the cheap production of simple shapes and thus give more individual components and assembly tasks. At the design stage, there is thus an opportunity to make assembly completely superfluous or at the very least to simplify it by selecting appropriate forms and fabrication methods.

In the preparation for production and the production itself, the designer's choice of method is supported by the choice of production sequence, equipment and the length of production run, and the utilisation of the equipment can be optimised. But the possibilities for rationalisation which are to be found here are modest seen in relation to the designer's influence.

Design for assembly[39] can be carried out at three levels:

- At the product-range level, where one attempts to fix on a strategy for the area of assembly and to get the product range to match this strategy, by for example establishing equipment which can assemble several different products, or which is flexible with respect to changes in the product. At this high level, choice of processes which make assembly superfluous can lead to dramatic reductions in costs.

- At the product-structure level, where one attempts to structure the product in a manner which is particularly well-adapted to assembly. The structure defines the components and their assembly methods and thus contains the seed from which the definition of the assembly task grows. A number of different structure types are advantageous with respect to assembly, and by thinking in terms of integration and differentiation of components the designer gets an overall view of the possibilities.

- At the component level, where one attempts to define the components so they most easily fit into and support the task of assembly. Here the designer must know and individually consider all the operations which are used in the assembly, and must for each of them attempt to match the components as closely as possible to the demands made by the operation.

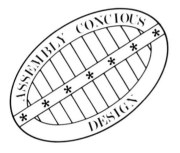

Design for assembly can be carried out in different ways. One can attempt to incorporate knowledge about production technique in a design review or a value analysis activity, and in this way revise the product. However, it is unfortunate to have to correct already finished drawings, so people with a knowledge of processes and assembly should therefore be called in to help with the actual definition of the product's structure and components.

The characteristic pattern of events which indicates when this should take place is just exactly the simultaneity of phases within the Integrated Product Development model. The production department can buy itself an assembly rationalisation by being an active participant in the creation of the product, i.e. the production department's interest in the product must have a very early starting point.

> Too many important things take place in the design phase for it to be left to the designer on his own

In certain cases, both the product and the equipment to assemble it are developed in the same project. This simultaneous development gives good opportunities for optimising both of them, and the integrated method means that the total development time is shortened, development costs are reduced, and better business can be created because business is chosen as the objective.

Attention to assembly during development of the product obliges the project leader to follow particular strategies. The following are provocative proposals[39]:

- Don't try to force product development. If both a new product and new assembly processes have to be developed in the course of a project (and the development team haven't tried it before), then time is required.

- Forcing the start of the project gives problems later. Faulty decisions in the early stages cannot be corrected by adjustments in the following ones. Careful general decisions about assembly must be made at the start.

- A short development time can be accepted if you at the same time plan the next generation of products. In the project for the next generation you can then take into account experience and ideas which couldn't be realised in the first generation, and you can correct the mistakes which were made. The risk involved in having to force the project is then reduced.

- The first assembly projects do not give a good return on investment; an extra effort must be expended if the staff have not attempted assembly rationalisation before. Additional costs in the design phase, in setting up production and in running-in can exceed all limits.

The rationalisation of assembly should not be considered as an activity which belongs to one product development project, but as

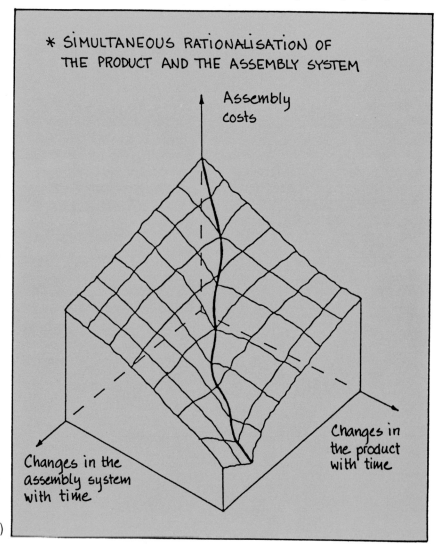

* SIMULTANEOUS RATIONALISATION OF THE PRODUCT AND THE ASSEMBLY SYSTEM

Assembly costs

Changes in the assembly system with time

Changes in the product with time

8.8 (after(40))

a contribution to a unified design and production strategy. As shown in Fig. 8.8, a gain can be obtained from rationalisation both through alterations to the product and through assembly automatisation, but the effect of doing both at once is considerably greater. A strategy like this requires us to map out the 'mountainous terrain' and to work out rather concrete targets for the area of design – the product which is ideal from an assembly point of view – and the area of production – the factory which is ideal from an assembly point of view[40].

8.4 Process quality

In the area of process technique, the same laws apply as in the area of assembly, and we shall only mention them in brief:

- Process-related costs are decided early and come into effect at a late stage during the development project.
- In the design phase, the designer decides, consciously or unconsciously, all important factors with respect to processes – i.e. their type, costs, efficiency, quality, and so on.

Just as in the area of assembly, it is an important problem to get know-how about production technique designed into the product in good time – i.e. at the time of definition of the structure and the components.

We have on many occasions in this book pointed out the type of unified view which is needed to attain sensible results (see the examples on pages 16, 107 and 110. Very often, a rapid concretisation and subdivision of the product into components takes place, and it is precisely subdivision or the creation of structure which is the key to substantial portions of costs.

The process or production area comprises three phases, which are radically different:

- The early design phase, where the designer's influence is enormous.
- The process specification phase, where the production engineer's professionalism leads to a complete definition of the method of fabrication.
- The implementation phase, where unforeseen problems with the production or with the quality of the product turn up and must be solved at a critical moment.

In all three phases, both design and production are involved. Good solutions appear when the designer continues with the project to the bitter end, and when the production engineer is engaged in the early activities. A good measure of the scope of this is the 'vanishing point for interest', i.e. the moment at which a project participant thinks that his or her task has come to an end. Notice the use of the word 'thinks'. The real influence of the participant, and the point at which it would be defensible for him or her to sever all connection with the project, reach much further out than the participant thinks!

The area of design for process is not so well-developed as it ought to be. People assume that the designer and product developer has a good knowledge of production technique, even if it was several years ago when he qualified as an engineer, even if the place where he got his degree had outdated equipment and only offered courses in traditional processes – and even if he only rarely turns up in the factory itself.

The early phase of specification is for the designer the early phase of design. As stated previously, a very considerable part of production costs are decided in this phase – up to 70% – and the important problem is to see a connection between design decisions and costs. A number of methods can be used:

- Relative costs: Tables are available which show the relative costs associated with the choice of various common solutions. It has been shown that relative costs are unaffected by changes in absolute costs, such as price increases for materials.

- Calculation programs: Interactive programs are available for the calculation of costs, where the designer replies to questions from the computer about the geometry of the object to be produced, and gets told about cost contributions and total costs.

- Delta calculations: Very often there are known designs which are very similar to the one currently being considered, so one can make do with looking at increases or reductions in costs which can be introduced in the current design.

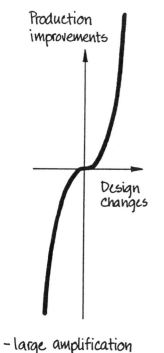

In the early phase of specification, there is no need for exact calculations of costs, but rather for rough indicators which can assist with the task of choosing between alternatives. As mentioned in Section 5.6, direct wage costs play an insignificant role, while materials and indirect costs are the dominant contribution. Thus the designer should use costing models which include these contributions and make it possible to handle them. A rough, but good, method is:

- Counting operations: When an object is defined, this gives rise to a series of operations in the production process, such as procurement, processing, checking and assembly. Counting the numbers of operations, and comparing them for two alternatives, can give a very interesting indication of the costs structure.

CAD/CAM is often given prominence as an important tool for the exploitation of knowledge about the costs involved in earlier designs, when making new designs. While this is undoubtedly true of CAD/CAM, there are also two dangers in it: that the model of costs be too limited (remember Fig. 5.30!), and that what the computer 'knows' be too conservative. The product model which the computer contains is based on a 'section' through the design process, where the product principles, structure and many of the

182

8.9

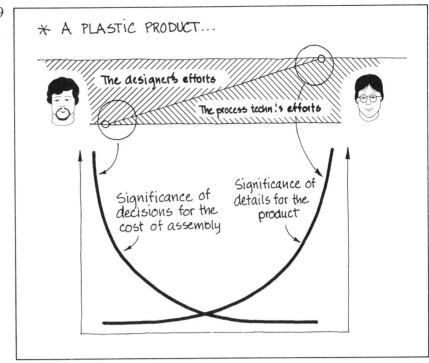

aspects of its detailed design have all been fixed, and therefore cannot be manipulated in the computer. A CAD system, if used for designing plastic-reinforced propellers, will never manage to tell you that laminated wood might be cheaper!

Insight into the processes used within the company and some operational numbers for production costs – what does a blind hole cost, a punched hole, grinding per metre, welding per metre? – is an important requirement for the designer who is to be able to take on the actual task of design. In addition, he should get used to 'talking to' people who are knowledgeable in the process area at the drafting stage – i.e. to working out drawings and sketches which are well-suited to communication. During the design of, for example, a small plastic object, even a very modest contribution during the early stages of fixing the design from one who knows about plastic and tools for plastic, can lead to good results. On the other hand, the designer should continue with the project right until the production is running as planned, as he with very little effort can give the production engineers just those degrees of freedom they need to solve their problems (Fig. 8.9).

We need a realistic view of the extent of the problems. The management has to learn:

When the prototype works, 80% of the work is still to be done!

The implementation phase is critical, because unexpected problems appear when the production has to be run-in and the product's quality requirements are to be met. The phase is critical because a shift of responsibility takes place; perhaps the project leader thinks the designer is finished, and that the production engineers need some small changes to the product to get things to

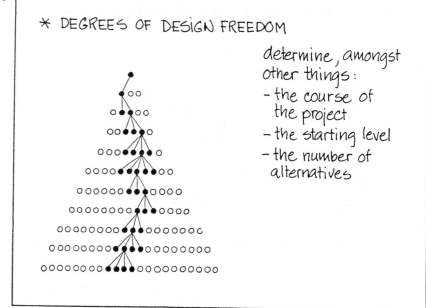

* DEGREES OF DESIGN FREEDOM

determine, amongst other things:
- the course of the project
- the starting level
- the number of alternatives

fall into place. In this phase, the quality of the company and the collaboration and competence of its employees are exposed. A direct measure of quality is the length of time in which a buck can be allowed to pass before 'somebody' looks after it[41].

8.5 Design methodology

Our dream in dealing with product development is to know all the strings which we can play on, in other words all the degrees of freedom. What kind of parameters get fixed in the 'market arrow' as time goes by, and how are they related to successful, high quality marketing? What kind of parameters determine technology, processes and production, and are essential for effective, flexible production with guarantees of delivery?

In the area of production we have a good suggestion for what the answer should be, and that is the design degrees of freedom presented in Section 4.3. These degrees of freedom give a picture of how the project proceeds. At every level there is a long series of alternatives, of which we can choose the best, which then becomes the starting point for the search for a solution at the next level (see Fig. 8.10).

What is design methodology? It is a coherent theory about methodology and the use of systematic methods in the creation of machines and mechanical products. We shall look more closely at its important features:

- General Problem Solving
- Product Synthesis
- Integrated Product Development
- Systematics

General Problem Solving (see Fig. 5.18) is a generally valid procedure for the solution of open problems, i.e. problems where there are many solutions[3]. In this procedure, the following ideas are used:

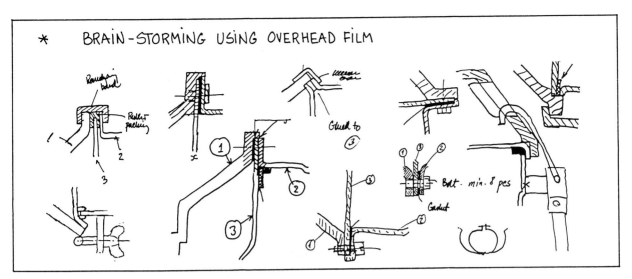

8.11 (after(3))

- Problems arise from a need, i.e. an unsatisfactory situation for 'somebody'. Problems don't formulate themselves, but are derived from the need situation on the basis of interpretations, objectives and policies – not made by those which have the need but by those who put themselves to solving the problem. This implies political horse trading when it is a question of building a large bridge, or company policies when it is question of interpreting market needs and transforming them into projects and products.

- Criteria for good solutions must be formulated before we search for the solutions, otherwise our criteria will be coloured by what we think is a brilliant solution. Dealing with requirements and criteria is discussed under Product Specifications in Section 6.7.

- The best solution (i.e. the best compromise) must be chosen on the basis of criteria, and must be made more concrete and definite, so that it can be communicated to those who will make use of it or realise it in practice.

- There is a relationship between the number of alternative solutions that we find and the quality of these solutions: There are more good ones if there are many. We cannot evaluate a single solution – our methods of evaluation only permit comparison – so we need alternative solutions in order to show whether it is likely that our brilliant solution is the best one.

A long series of methods for dealing with the perception of need, formulation of problems, evaluation and so on are associated with this way of going about things. Even more important, of course, are methods for searching for solutions, the so-called creative methods. These are methods which exploit various mechanisms in our way of thinking. Figure 8.11 shows the result of a brainstorming session in which each participant was given some pieces of overhead film to formulate a proposal for a solution on. On the overhead projector, all the proposals can be seen at once, one can discuss them, sort them, choose between them, and perhaps even carry out a new round using better principles on which to base the solution.

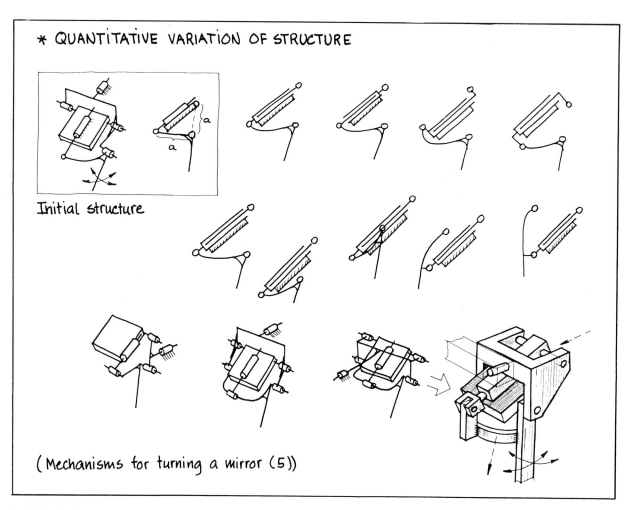

* QUANTITATIVE VARIATION OF STRUCTURE

Initial structure

(Mechanisms for turning a mirror (5))

8.12 (after(5))

8.13 (after(39))

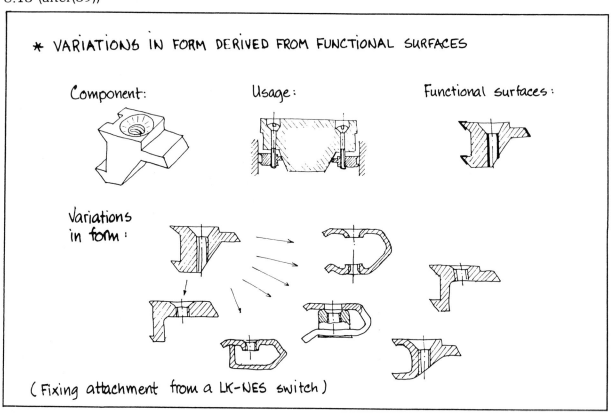

* VARIATIONS IN FORM DERIVED FROM FUNCTIONAL SURFACES

Component: Usage: Functional surfaces:

Variations
in form:

(Fixing attachment from a LK-NES switch)

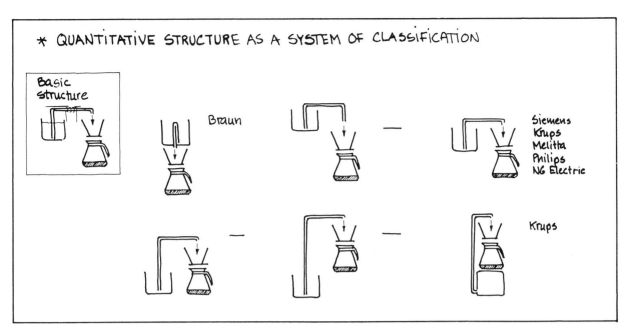

8.14 (after(5))

Product Synthesis (see Fig. 6.1) is a design procedure for machines and mechanical products. It is based on the theory of design degrees of freedom, and specifies a sequence in which a machine is defined: process, function, structure, form, etc. All machines' or products' properties are dependent on and defined by the so-called basic properties of structure, form, materials, dimensions, surface quality and tolerances. It is just exactly these things that we specify on a workshop drawing. A very large number of methods are associated with product synthesis. These methods take as their starting points laws governing the nature of machines – e.g. that a machine's structure can be varied quantitatively, and that this quantitative variation leads to alternatives, as shown in Fig. 8.12, or that a mechanical part's function is defined by its functional surfaces and that one can therefore create new shapes by varying the so-called free surfaces (see Fig. 8.13).

Integrated Product Development is not a constituent of design methodology, but forms a framework for design activity, and must therefore respect the nature of the design task. One of the starting points for setting up the model for Integrated Product Development in Chapter 3 was the area of design methodology, which thus forms a basis for rational product development. Thus companies should also take an interest in the way in which their staff actually design things. Design methodology forms a good pattern for design work, into which a long series of technical areas such as dimensioning, choice of materials, control technique, ergonomics, product liability, machine elements, CAD, design, mechanisms, process technique, and so on, can be incorporated and made operational.

Systematics and systematisation make up an important element in the area of design methodology. Linnaeus created a systematic classification for flowers and Darwin for animals, the Patent Office a classification system for patents, and the libraries a decimal classification system for subjects – in the same way we can for a particular product area create a universal description of all product alternatives which can solve the current problem. It is of course an

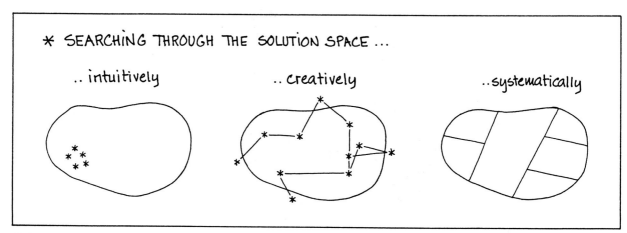

* SEARCHING THROUGH THE SOLUTION SPACE ...

.. intuitively   .. creatively   ..systematically

8.15

idealised and unrealistic idea, that one should find 'all' solutions in a development project, but it can have considerable significance at a more general level to map out the principles which can be used and the possible ways in which a product can be built up. An overall view of this type can not only serve to indicate new principles, but also to map out principles subject to patents and principles in use by the competitors (see Fig. 8.14).

Methodology and systematics are measures of how well the available solution space is covered. The space is determined by what we have to attain in relation to our competitors, i.e. our level of ambition and the risk we are willing to run. We create the boundaries of the solution space when we formulate the BS and PS for a project, and the methods which are used define how well the solution space is searched for good solutions, and how well we map out where our competitors are to be found (Fig. 8.15).

The integrated company

SUMMARY

The main subject of this book is the need for integration in development projects, and we have seen various methods of attack and tools which can help with the solution of the many problems which arise in the specialised and divided company.

In this chapter the claim is advanced that a company's numerous streams of money, materials and information are created in its development projects, and must thus be rationalised and controlled by intervention in these development projects.

The company which is concerned with renewal must regard it as its task to perform integrated control of the most important cost parameters of the day-to-day running of the company, through the choices which are made as part of its development activities.

9.1 Development projects

Product development is associated with a considerable element of risk, as unforeseen things can occur in the world outside (the market can collapse, strong competitors can launch a new product in front of your nose) or in the project itself (the interpretation of the market was faulty, the chosen principles gave quality problems, the customer didn't like the product). These matters are impossible to control.

Product development can be compared to sailing in a sailing ship. Regardless of your preparations, equipment and seamanship, you need wind in your sails in order to sail away.

The output from creating business cannot be controlled

Input control means that you do what you believe to be the right things, and create the best possible conditions, so that the probability of attaining a good result is maximised. But you have no direct influence on whether the business operation will succeed.

A company can be compared to a ship, as far as its day-to-day running is concerned. An unseaworthy coffin ship, poorly trimmed, with poor sails badly set, and incorrectly navigated, will sail much less efficiently than is possible. In the same way, there are a large number of quantities within a company which affect the efficiency and profitability of its day-to-day running. The efficiency can be measured where results are created and where things 'flow' in the company, so we must take an interest in flows such as:

- Flows of money, such as sales income versus material costs, wage costs, capital costs and so on.
- Flows of goods, i.e. raw materials, semi-manufactures and components flowing from subcontractors through the factory into intermediate and final storage and eventually to storage at the wholesaler's, all of which are checked and transported several times.

- Flows of decisions and realisations, i.e. sequences involving the recognition of facts, diagnoses, panels, consequence analyses, decisions, the formulation of targets, setting things going, following things up, evaluation of results.

- Flows of information, from the world outside to the company, from the management to the staff, from previous experience to use in new products, from the marketing to the development department, within the production system.

- Flows of paperwork, such as the papers necessary for defining, purchasing, controlling quality, controlling materials, controlling processing, assembly and test of the components which go to make up the product.

Together, these flows describe the day-to-day running of the company and the costs which are associated with it. They turn up in auxiliary functions and in auxiliary functions for auxiliary functions. Thus these flows can be considered as derived quantities or auxiliary quantities; for these it is the case that:

> All derived quantities in the creation of business can be controlled!

To use the image from before: we cannot ensure that we get wind in our sails, but we can effect and organise all other factors so the wind has the greatest possible effect.

The relationships within a company, between the many types of flow and the parameters which describe the results, are very complex. A suitable image is a cobweb, whose radial threads are important parameters. If we change a parameter or an auxiliary quantity, corresponding to our pulling in one of the threads, then the whole cobweb gets distorted. Our problem is to detect whether what we have done has some influence on the desired parameters, what sign this influence might have, whether it is worth the trouble, and whether it has unfortunate consequences somewhere in the system.

But at least there is one place where with some certainty we can affect a large number of the parameters and obtain a considerable effect – and that is in product development. All the flows and other conditions within a company can be seen as a consequence of the decisions which were made at the time when the business operation

9.1

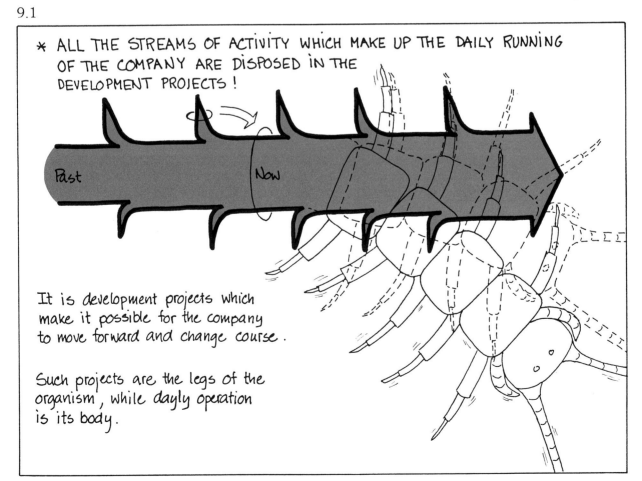

* ALL THE STREAMS OF ACTIVITY WHICH MAKE UP THE DAILY RUNNING OF THE COMPANY ARE DISPOSED IN THE DEVELOPMENT PROJECTS !

Past   Now

It is development projects which make it possible for the company to move forward and change course.

Such projects are the legs of the organism, while dayly operation is its body.

was being set up; here we are thinking of the sales system and the production system, while the structure of the product defines the flows of goods, of money, of information, and so on. In each individual component lie the seeds for the flow of information – to the supplier of materials, to the quality control department, through production, assembly, checking, etc. – and for the flow of materials. Each individual flow is associated with costs and is thus an area which can be tackled with the tools of rationalisation.

Figure 9.1 is an illustration of this. The horizontal arrow symbolises the flow which, taken all together, makes up the day-to-day business, while the small vertical contributions to it are the development activities. Each individual project contributes to the overall business, and at the same time contributes to the nature, sequence and costs of the total flow. The flowing quantities are dependent on parameters of the product, its production and its marketing, so that:

A company without product development cannot be controlled – it cannot alter course

9.2 Control of costs

We noticed in Section 5.6 the paradoxical feature that we try to rationalise in a manner quite out of proportion with the consequences associated with the various contributions to costs (Fig. 5.30). Unfortunately, our detailed knowledge of costs is greatest in those areas where it has no significance, and very superficial in those areas where control would give an advantage. Accountants don't bother about the areas of technological costs, i.e. materials and capacity costs.

What we need is costing models and insight into costs which, described in terms of the cobweb model, can tell us which consequences a change will have in those parts of the cobweb where an advantage is to be had. The relationships which exist (Fig. 9.1), between our decisions in the early phases of development and the financial structure of the day-to-day running of the business, are so complex that we should be happy if we can do as much as determine the sign of the consequences of what we are doing. In development, we need to think about marginal changes: 'What happens if we do this somewhat differently?' The return-on-expenditure method of describing costs is not valid in connection with delta considerations during the development phase, because it is here we manipulate the return on expenditure.

Companies can normally be run in many ways – all at once! Normally, you control costs by controlling production, but it is also possible to let the 'designer' control the product. The 'designer' is here to be understood as being those people who took part in the early phases of the design. Those decisions which are made at that time are irreversible!

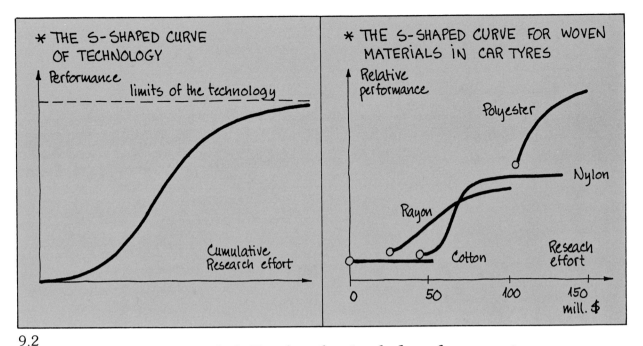

9.2

9.3 Technological development

Controlled development within a company requires the formulation of objectives and strategies. Each of the three areas – market, product and production – require long-term strategic targets, if the company is to be made controllable in each of its development projects.

Technology, in the sense of 'how things are done', is a good high-level term for the means which can be chosen for realising the three types of strategy. One can speak of marketing, product and production technologies, and technologies have to be looked for and developed. Product development is about transforming technology into business. In this respect, every technology gives a yield which follows an S-curve, as shown in Fig. 9.2A. The transition to a new technology, a so-called technology jump, can radically change the yield of a technology, as shown in Fig. 9.2B.

9.3

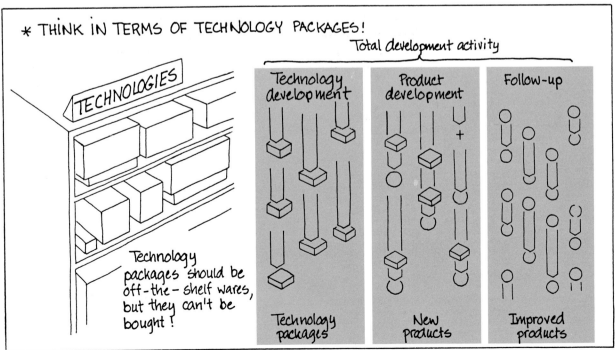

It costs money to creep up along the curve of technological development until you reach the middle part of the S-curve, where the returns start to be good. Thus technological development must be carried out with care. A common pattern is to carry out technological development within the framework of a product development project. However, there is much that indicates that it should rather take place as independent activities and lead to off-the-shelf technology, so-called 'technology packages' (Fig. 9.3), so that the development of a product is not adversely affected by the uncertain activity of creating a new technology. The development of technology packages should be project-oriented, but with far-sightedness, willingness to take risks, and using experimental procedures, when it is a question of technologies which are decisive for the company. Remember that what we are selling is technology.

It is important to recognise your technological basis and consciously to direct things towards strengthening important and profitable technologies and suppressing or replacing weak technologies. It can be extraordinarily difficult to foresee market requirements in 10-15 years' time, and thus in certain companies one can bank on technological development and just see what this leads to.

Technological development, as a fuel for product development within the company, is of vital importance for our future. Thus companies should take steps to find out who has responsibility for the development of individual technologies, and must formulate strategies which ensure that these technologies contribute to the competitiveness of the company in the long term.

9.4 The ideal factory

The situation in the area of production is today filled with problems. We are in a transition period, where we have to leave behind us the production systems which were created out of a stable situation 10-15 years ago, and instead create production systems for new tasks. The task of production today is characterised by frequent changes, uncertain market conditions, increased demands on quality, and desires for automatisation and computerisation in the organisation and control of production.

9.4

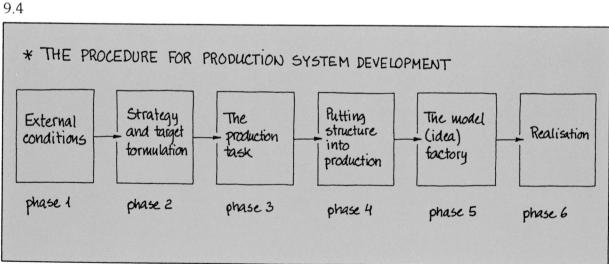

Production has up till now been developed in small steps, which have not always been well coordinated. These days, renewal must take place in big jumps, if we are to keep up with the development which is taking place in the leading industrial nations.

At the Institute for Product Development, the Management Theory Group has created a significant new way of going about things for use in the Development of Production Systems, which we call UPS[9]. The basic idea is the concretisation of the task of production, and the formulation of a concrete target, the 'ideal factory', for the development of the necessary production technique. What can be tackled using the UPS procedure is production targets, the structure and control of production, production processes (technology), and the organisation of production, together with its associated types of collaboration.

The 'ideal factory' is based on clear objectives with a high level of ambition, so as to create a leap in development, and attention is focused on the contribution made by production to external efficiency, i.e. to the company's exploitation of possibilities presented from outside. Thus it is features such as flexibility, adaptability, ability to develop, quality and precision in meeting delivery dates that are drawn to the fore, rather than productivity.

The UPS procedure is illustrated in Fig. 9.4. It bears a great similarity to the product development model of Fig. 4.5; however, it does not cover a single product-related activity but the sum of a number of existing and future production tasks, within a timescale of 3-5 years. The UPS procedure's primary focus is on the need for the production department to solve the production development task. Stress is laid on the fact that you must consider both the new production tasks which arise in each individual development project, and the overall production task (see Fig. 9.5). But the products are considered as random input, which cannot be influenced.

In the design phase, up to 70% of costs, including a considerable part of the production-related costs, are tied up. The processes and the properties of the production are to a great extent fixed by the designer's drawings:

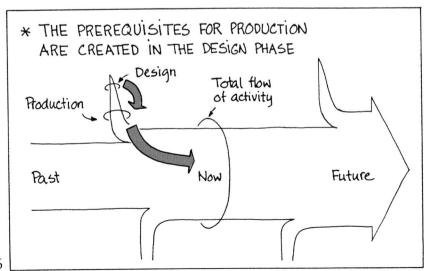

9.5

We have seen previously (e.g. in Fig. 8.6) how the area of design is a very important partner in the realisation of production strategies. Thus the preceding project activities marked in Fig. 9.5 should be considered as the adjustable and most important part of the foundation for the production activities within a project, while these in their turn form the basis for the overall flow of activities (cf. Section 9.1).

9.5 The ideal product

The situation within the area of product development is, as indicated in the introduction to this book, surprisingly difficult. The increased competition forces us to develop competitive products, which at the same time improve the company's cost structure. The prerequisites for the development task, that is to say the task of developing and designing new products, have also changed: we now have new insight into the significance of the task of design – amongst other things, as presented in this book – and we have new tools, not least computerised systems, to help with the solution of the more or less 'global' tasks within the company.

The development task can naturally be considered as the creation of that product which gives good business, but it can also be considered in terms of the development group's tasks, their qualifications and objectives, such as:

- To supply the company with new products which can maintain/expand the basis of its business.
- To keep up the company's product range, both from the point of view of technology and with respect to the market.
- To establish and maintain a development organisation which corresponds to the results of the development task and the resources of the company.

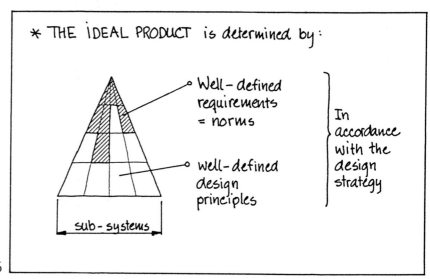

9.6

196

- To demonstrate technological preparedness, i.e. to ensure that the company is up to the mark – perhaps even a bit in front – and has a flexible attitude to new technologies.
- To demonstrate precision in their perception of needs and markets.
- To demonstrate results in the form of competitive products.
- To create products which increase the company's internal efficiency, i.e. ones which are optimal with respect amongst other things to finance, materials, subcontractors, marketing and production.

Just as in the UPS procedure, one can attempt to identify and make concrete the ideal product. This is not to be understood as a good product business-wise, but as the product which – as mentioned in Section 9.1 – is ideal with respect to all the derived quantities, such as its cost structure, component structure, subcontractor structure, quality structure and service structure, to name but the most important.

The ideal product should be a very concrete objective, i.e. a clear description of which design solutions we can see will lead us to the realisation of our development objectives, and which requirements must be considered as given and fixed. These considerations can be expressed in terms of the design degrees of freedom, as shown in Fig. 9.6.

There is an area which must be considered as a necessary requirement for every project, and which must not be tampered with: it can be the product's area of use, its principles of operation, particular features of its construction, materials, the choice of process, and so on. How much is fixed depends on the nature of the individual subsystem, and on the level of ambition of the development task.

Apart from its fixed features, the ideal product can be defined using design principles, i.e. solutions which support the strategic objectives which have to be fulfilled in the development task. This might be a requirement: 'use sheet metal' in order to be able to use a universal raw material and to utilise sheet metal technology; or it might be 'design for the use of complex plastic tools' so aspects of the assembly are improved, or 'use module X from product Y' in order to affect degree of re-use, length of production run, storage size and so on, or 'make parts exposed to wear easily accessible' so that service or repair by the customer himself is made easier, or 'integration of hydraulic parts' so fittings and pipes are eliminated, etc.

Definition of the long-term objectives for the development task in the form of the 'ideal product' can be considered as the elaboration of standards. These standards must partly ensure that the task of development does not tamper with the established foundation which, perhaps over a long period, has led to the company's position of strength. And they must partly ensure that the results of the project are directed onto, and contribute to, the line which the company wants to follow.

It is through more or less conscious work with the establishment of standards that the Head of Design controls design activity, so its contributes to the efficiency of the company. We cannot afford to develop products which affect the quantities that flow within a company in random directions; we must control innovation, so the derived quantities are under control. But the business itself must not get tied down.

9.6 Integration

How do we find the right balance in development between, on the one hand, giving the project a free rein, so that results with a high degree of innovation are created, when we, on the other hand, know that the design of the product has its influence on every aspect of the company, and thus needs control? Part of the answer is that we must see the connection between the development projects, isolate the 'conservative' portion of them and then increase our efforts as far as possible towards the innovative targets.

This is only one of the many dimensions of integration which are to be found within a composite company with many complex development projects. In the first chapter of this book, we have indicated those areas of integration where we believe that a job needs to be done with the glue, i.e. the tools for integration, if we are to have any hope of there being any coherence in the company's activities.

We have in this book presented the concept of integration in connection with product development, and have in this way made a contribution to the long list of empty concepts which are so easy to manipulate, and which so easily get to manipulate us: system, flexible, universal, organisational, CAD, FMS, CIM, etc. Just as the concept of flexibility demands further explanation of 'relative to what?', so integration requires a further specification of 'in what respect?' and 'of what?' For example, integration between design and production with the aim of controlling materials costs.

We have not been able to sharpen the concept in this way in this book, because the area is a new one, not everything is yet mapped out, the tools have not yet been fully identified, and because it would make for heavy reading.

We have attempted to relate the mechanisms of integration to the nature of innovation, product development, design and problem solving, and believe that we in this way are able to create more radical procedures than ones which are based on peoples' behaviour and interactions.

* PRODUCT DEVELOPMENT IS REVOLUTION

9.7

But, true enough, it is a planned and controlled revolution. It is revolution for those who have been put to look after the day-to-day running of things, where all changes are disturbances (Fig. 9.7). So you have to make use of both the conservative and the innovative elements in the company, in recognition of the fact that innovation is the legs of the animal, but that the animal has no body if its day-to-day business does not take place.

9.7 Development tendencies – An example

In recent years, development work has become very heavily affected by the possibilities offered by computers[42]. The computer is considered as an integration tool, which leads to direct information routes from customer orders and market development via planning, materials control, stores control and the creation of a basis for production, over to the control of goods and production. Or which leads directly from the designer's sketchbook via dimensioning, optimisation, the preparation of drawings and the preparation of process control data to running production. Or which leads us from experience in earlier projects, experience with technologies and production costs and so on, to better decisions in new situations of choice.

At present we have to make do with relatively trivial views of the whole in the so-called integrated systems. We can handle many elements in the computer, and in certain areas we can connect these elements to one another, but it is still reasonable to compare such systems with a box full of unsorted pieces for a jigsaw.

Many companies are currently gaining experience with CAD systems, principally systems which permit the production of drawings. These companies discover that:

CAD is revolution!

The computer obliges one to make a long series of decisions, and requires one to follow a well-ordered sequence of design, in contrast to manual design, where the flexibility of the human designer permits everything, and can easily handle inconsistency, incompleteness, preliminary data, and so on...

The order introduced into design work is just exactly the standardisation of its starting requirements mentioned above. For most companies, an enormous effort is required to keep track of their own wildly expanding product range, and to create the archetype or 'model product', which is to form the basis of future designs.

The paradoxical thing is that when the result of this standardisation becomes available, then the company recognises how incredibly valuable it is, but at the same time the company also recognises that the task could easily have been carried out 5-10 years ago, and would not have required tools other than a copying machine, a pair of scissors and some glue. The computer obliges us to standardise, but it is integrated cost control which gets the main benefit of this, and which should be the main objective of the exercise!

Prospects for the future are of course not completely covered by a discussion of CAD/CAM, CIM or other computer-oriented concepts; but with this excursion into the area we wanted to demonstrate that we consider it an important task to find tools which permit integrated manipulation and control of the most important cost parameters in the day-to-day running of the company, through the choices which are made during product development projects. We must be innovative and creative in our way of creating business, but all the decisions which we make about the realisation of products must be tightly controlled, so that the company moves toward an attractive future.

9.8 Concluding remarks

In this book we have dealt with professionalism in product development and with those mechanisms and tools which can counteract the effects of division of tasks and specialisation in a company. The integration concept affects both the individual project and the interaction between projects and the development of the company.

This book has been written for readers who are aware of what product development is, and who are involved in it. But we believe that the book will also give readers who have other functions or qualifications a coherent picture of the relevant ends and means, despite the book's variegated and essay-like form.

It was not our wish to present the integration concept or the creation of the 'ideal product' as a fixed procedure. Experience with the use of integrated product development in industry is still too limited for that. But the feedback which we have received has been strongly favourable.

What should the reader do now? This book must be seen as a contribution to the continual process of modernisation which takes place in industry. The reader must recognise that many things which were once considered as unchangeable, and which led to particular rules and patterns of behaviour, have now changed so much that we have to revise our methods and ways of going about things.

<u>Integrated Product Development</u> indicates some of the changes required.

References

[1] Eckels, J. Lecture notes. Danfoss, 1983. See also, Eckels, J. Industriële Doelontwikkeling (dissertation). Delftse Universitaire Pers, 1982.

[2] Freudendahl, J. and Bonde, H. Restructuring of Agricultural Trailer Programme (in Danish), Final year project. Institute for Engineering Design, DTH, 1983.

[3] Lecture notes series, Problem Solving (in Danish). Institute for Engineering Design, DTH, 1985.

[4] Hubka, V. Theorie Technischer Systeme Zweite Auflage. Springer-Verlag, Berlin, 1984.

[5] Tjalve, E. A Short Course in Industrial Design. Newnes-Butterworth, London, 1979.

[6] Andreasen, M. M. Machine Design Methods Based on a Systematic Approach (in Danish), Dissertation. Lunds Tekniska Högskola, 1980.

[7] Freudendahl, J. and Bonde, H. Purchasing (in Danish), IPU note 026-K-85. Institute for Product Development (IPU), Lyngby, 1985.

[8] Product Development and Design (in Swedish). Sveriges Mekanförbund, Stockholm, 1980.

[9] Development of Production Systems (in Danish), Reports. Institute for Product Development (IPU), Lyngby, 1983.

[10] Miles, R. E. and Snow, C. C. Organizational Strategy, Structure and Process. McGraw-Hill, New York, 1978.

[11] Mikkelsen, H. The Project Team Member (in Danish). Promet, Vedbaek, 1984.

[12] Roth, K. Konstruieren mit Konstruktionskatalogen, Springer-Verlag, Berlin, 1982.

[13] Olsson, F. Systematic Design (in Swedish), Dissertation. Lunds Tekniska Högskola, 1976.

[14] Clausen, J. Company Goals (in Danish), Lecture notes. Danfoss, 1984.

[15] Mikkelsen, H. and Riis, J. O. Basic Project Management (in Danish). Promet, Vedbaek, 1985.

[16] Rothweller, R. and Gardiner, P. The Role of Design in Competitiveness. Design Policy: 2. Design and Industry. The Design Council, Royal College of Art, London, 1984.

[17] New Products Management for the 1980s. Booz Allen & Hamilton Inc., New York, 1982.

[18] Haasen, S. Integration in Companies (in Danish). Institute for Product Development (IPU), Lyngby, 1986.

[19] Systematische Produktplannung, WIFI 115, Düsseldorf, 1980.

[20] Develop New Products! Study material. Sveriges Mekanförbund, Stockholm.

[21] Harris, J. S. The new product profile chart: Selecting and appraising new projects. Chemical and Engineering News, Vol. 39, No. 16, 1961.

[22] Hubka, V. Principles of Engineering Design. Butterworth Scientific, 1980.

[23] Innovation or How to Make Things Happen, Technology review. MIT, Cambridge, Massachusetts, 1976.

[24] Andreasen, M. M. and Hein, L. Travelling report – Japan 1984 (in Danish), IPU note 032-K-85. Institute for Product Development (IPU), Lyngby, 1982.

[25] Hein, L. and Pedersen, F. M. Travelling report – USA 1981 (in Danish), IPU note 006-K-82. Institute for Product Development (IPU), Lyngby, 1982.

[26] Folman, E. Notes on Project Culture (in Danish). Projektplan, Teknisk Forlag, 1984.

[27] Andersen, S. K. Tracing the Costs (in Danish), IPU note 023-K-84. Institute for Product Development (IPU), Lyngby, 1984.

[28] It's not so much what the Japanese do . . . it's what we don't do. Ingersoll Engineers, Rugby, UK, 1982.

[29] Freudendahl, J. and Bonde, H. Catching the Costs (in Danish), IPU note 027-K-85. Institute for Product Development (IPU), Lyngby, 1985.

[30] Flurscheim, C. Engineering Design Interfaces: A Management Philosophy. The Design Council, Royal College of Art, London, 1977.

[31] Valbjoern, K. The Early Phases of a Design Project (in Danish), Project No. 1, 1982.

[32] Andersen, S. K. Planning with Key Points (in Danish), IPU note 033-K-85. Institute for Product Development (IPU), Lyngby, 1985.

[33] Fangel, M. Project Management (in Danish). Projektplan, Teknisk Forlag, 1982.

[34] Risk Assesment series. Mechanical Engineering, November 1984.

[35] Casher, J. D. How to control project risk and effectively reduce the chance of failure. Management Review, June 1984.

[36] Hartz, O. Total Quality Control (in Danish). Danske Ingenioeres Efteruddannelse, Lyngby, 1977.

[37] Schleischer, E. Quality Strategy, Lecture notes. Danfoss, 1984.

[38] Product Development (in Swedish), Course lecture notes. Alfa Laval, 1982.

[39] Andreasen, M. M., Kähler, S. and Lund, T. Design for Assembly. IFS (Publications) Ltd, Bedford, UK, 1983.

[40] Kähler, S. and Ahm, T. Design for assembly – A case study. In, Proc. 5th Int. Conf. on Assembly Automation. IFS (Publications) Ltd, Bedford, UK, 1984.

[41] Meyer, K. Starting New Products in Production (in Danish), Final year project. Institute for Engineering Design (DTH), 1984.

[42] The Strategic Benefits of Computer-Integrated Manufacturing. Arthur D. Little Inc., 1983.

[43] Buur, J. Integrated PVDF Handset for Telephones (in Danish), Final year project. Institute for Engineering Design (DTH), 1984.

[44] Spork, A. Coin Sorter System for Telephones (in Danish), Final year project. Institute for Engineering Design (DTH), 1984.

Bibliography/Further Reading

Brankamp, K. Plannung und Entvicklung neuer Produkte, De Gruyter, Berlin, 1971.

Buggie, F. D. New Product Development Strategies. Amacom, New York, 1981.

Design and the Economy. Report from the Design Council, London, 1983.

Flurscheim, C. Industrial Design and Engineering. Design Council, London, 1985.

Gerstenfeld, A. Effective Management of Research and Development. Addison Wesley, Reading, 1970.

Giragosian, N. H. (Ed.) Successful Product and Business Development. Marcel Dekker, New York, 1978.

Holt, K., Geschka, H. and Peterlongo, G. Need Assessment. John Wiley, Chichester, UK, 1984.

Leech, D. J. Management of Engineering Design. Wiley, London, 1972.

Pahl, G. and Beitz, W. Engineering Design in Theory and Practice. Design Council, London, 1986.

Urban, G. L. and Hauser, J. R. Design and Marketing of New Products. Prentice-Hall, Englewood Cliffs, New Jersey, 1980.